Leatherhead Food International

ESSENTIAL GUIDE TO FOOD ADDITIVES

Third Edition

Revised by
Victoria Emerton and Eugenia Choi

This edition first published 2008 by
Leatherhead Publishing
a division of
Leatherhead Food International Ltd
Randalls Road, Leatherhead, Surrey KT22 7RY, UK

and

Royal Society of Chemistry
Thomas Graham House, Science Park, Milton Road,
Cambridge, CB4 0WF, UK
URL: http://www.rsc.org
Registered Charity No. 207890

Third Edition 2008
ISBN No: 978-1-905224-50-0

A catalogue record of this book is available from the British Library

Typeset by Alison Turner.
Printed and bound by Cpod.

FOREWORD

The first edition of the Essential Guide to Food Additives was published in 2000, and its success merited the production of a second edition, which was published in 2003. The book continues to be a valuable reference tool for the food industry into the applications and legal restrictions governing food additives.

Since 2003, there have been significant developments in the additives area, and this third edition sees the addition to the book of 12 new additives (including two sweeteners) and the removal of several additives such as Red 2G (E128), propyl p-hydroxybenzoate (E216) and sodium propyl p-hydroxybenzoate (E217), which are now banned from use in food due to issues with their safety.

Biphenyl (E230), orthophenyl phenol (E231) and sodium orthophenyl phenol (E232) have been removed from the additives listing, as they are no longer classed as food additives, but are now plant protection products.

A general discussion on the use of food additives is provided, with the addition of a section on the new trend in clean labelling, and how additives can be used to best effect to fit consumer demand for clean labels. The General Food Regulations 2004 are included which complement the Food Safety Act of 1990 and further sections within this new edition cover the genetically modified food and feed regulations and information about the food improvement agents package.

As with any book containing legislation, every effort has been made to ensure that the information provided is correct at the time of going to print, but this is a fast-moving area, and readers are advised to check the current legislation on specific additives.

Acknowledgements:

The first edition of the Essential Guide to Food Additives was edited by Mike Saltmarsh of Inglehurst Foods Ltd, who helped to conceive the overall content of the book, and wrote many of the individual entries on additives, with other chapters and additives sections being written by experts from the food industry. Their contribution to the first edition is gratefully acknowledged. This third edition, like the second edition, was largely updated by the Regulatory Services Department, and members of the Scientific and Technical Information Unit at Leatherhead Food International. Particular thanks go to Eugenia Choi who has led the project and also to Katherine Veal, Claire King and Catherine Hill for their support.

Victoria Emerton and Eugenia Choi
Editors

CONTENTS

1. FOOD ADDITIVES AND WHY THEY ARE USED

Introduction

The role of food additives in food manufacture has been much maligned and misunderstood in recent years. Additives fell victim to bad press to the extent that, at the height of the anti-"E" numbers campaign in the 1980s, the word "additive" became almost synonymous with "adulteration", and foods containing additives were as much to be avoided as foods containing genetically modified ingredients have become since their introduction in the late 1990s. Authors whose main objective appeared to be the denigration of the food manufacturing industry, particularly the major multinationals, found this easy meat in an atmosphere of consumer ignorance, and were guaranteed support for their cause by scaring their audience into believing that additives were responsible for a wide range of ill effects from intolerance and hyperactivity to long-term chronic diseases. Constantly prefacing the words "food additive" with "chemical" was sufficiently emotive to result in the perception of "nasty". Alongside this was the implication that ready-prepared, processed food was inherently inferior to, and less wholesome than food prepared in the home.

The catalyst for the 1980s focus on additives was a change in labelling legislation in 1986, which required the detailing of each individual additive in the ingredients list of most pre-packed products. Until that time, the use of additives had been indicated by reference to a generic functional group, such as "preservatives", "antioxidants" and "colours". The new labelling requirements resulted in the appearance on some food labels of some very long lists of additives, including some lengthy chemical names. Some products looked as though they were nothing more than a couple of simple ingredients held together by a dictionary of chemical substances. The "E" number system, intended to assist as a short code for some of the lengthier chemical names and to indicate common European safety approval, became the butt of the criticism against the use of additives, and consumers voted with their feet by leaving products containing long lists of "E" numbers on the shelf.

The interest in, and fear of, what was being put into food spawned a number of books on additives, their use in food, potential (harmful) effects and protocols for their safety approval, along with the author's specific treatise on the subject. Some were informative, intended to assist the consumer in understanding what additives were, how they were produced, why they were used and how to avoid them, if desired. Others were more politically motivated and used the fashionable attack on additives as an illustration of all that was bad about the food industry and the allegedly secretive systems of safety assessment of all chemicals and processes used in food production. The implication was that any chemicals added to food, either as pesticides in primary production or additives in processing, were suspect.

A generation brought up on convenience foods, removed from the messy business of primary food production, fell easy prey to this suggestion, apparently oblivious to the substances and techniques employed by their grandmothers, when no self-respecting household would have been without baking powder, bicarbonate of soda, cream of tartar, a selection of flavourings and a bottle of cochineal – some of the most common everyday "food additives". These everyday ingredients might well be frowned upon by many a modern shopper uninitiated in the art of cookery, if spotted on the ingredients label of a manufactured product in the form of an "E number" or prefaced, as legislation requires, by its additive class. How many people think of additives when they buy a lemon or a bottle of vinegar? Yet these too are authorised additives (as citric and acetic acid, respectively) and widely used in food manufacture for their preservation properties, as well as their acidic taste, precisely as they are used in everyday cooking. The use of saltpetre as a preservative can be traced back to Roman times, and the controversy over additives use goes back to at least 1925, when the use of boric acid in food was banned under the Preservatives Regulations. However, in recent years the use of boric acid has been accepted under the Miscellaneous Food Additives Regulations 1995 as amended, but only for the treatment of caviar.

Whilst its complexity and scale do not lend modern food manufacture entirely to direct comparison with the traditional kitchen, it is often forgotten that the overall purpose is the same – to prepare, preserve, process and, as the case may be, cook basic raw ingredients to convert them into wholesome, attractive, better tasting and nutritious food, ready to be consumed. Every cook has his or her own techniques, and knows many a trick to prevent peeled vegetables and apples from browning, thicken sauces, brown the gravy, and transform an everyday dish into something special; he or she will also ease dinner party preparations by preparing in advance and storing the part-ready dishes for last-minute completion. Food manufacturers do much the same, and, over years of product development, first on the basis of trial and error and now underpinned by research programmes, have developed the most effective and economical methods of producing a wide range of foods to suit every taste and pocket. In order to achieve this, they need at their disposal a wide range of additives to perform a number of tasks in the process, from cleaning and refining the raw materials, to preserving them in optimal condition throughout further processing or distribution, combining them with other ingredients and ensuring that they appear attractive to the consumer. The types of additive used and some of the functions they perform are explored in greater detail below.

The anti-additives campaign and consequent consumer pressure to remove or minimise the use of additives inevitably led to changes in manufacturing practice and marketing. In addition, trends towards more "fresh" foods and the growth in market share of chilled foods, together with changes in legislation following completion of the European harmonisation exercise, all had an impact on the use of additives. It is therefore timely to review the place and use of additives in the food supply, whilst bearing in mind that they will always be essential to food preparation, quality and preservation.

What are Food Additives and why are they Used?

The use of food additives is nothing new. Preserving food is an age-old necessity. Many of the techniques that we now take for granted, such as canning, refrigeration and freezing, are relatively new. Even the overwintering of farm animals was rare until the 17th century, when feeding and husbandry techniques became better understood. Any old or weakly livestock such as oxen, cows, sheep, pigs and poultry had to be slaughtered in the autumn, and the meat was dried, salted or pickled to preserve it for the winter months (1). When food shortage ceases to be a problem, greater emphasis is placed on making food look and taste good, and we look beyond food as a survival necessity to food as a pleasure and a treat.

Food additives are used either to facilitate or complement a wide variety of production methods in the modern food supply. Their two most basic functions are that they either make food safer by preserving it from bacteria and preventing oxidation and other chemical changes, or they make food look or taste better or feel more pleasing in the mouth.

The use of additives in food preservation is, not surprisingly, one of the oldest traditions. Our forbears may not have thought of saltpetre, used as a curing agent, or vinegar (acetic acid) as additives, but they would have been the mainstay for ensuring a longer-term supply of precious perishable foods. Salt, though not an additive by the modern definition, was the other essential.

Food additives are defined in European legislation as "any substance not normally consumed as a food in itself and not normally used as a characteristic ingredient of food, whether or not it has nutritive value, the intentional addition of which to a food for a technological purpose in the manufacture, processing, preparation, treatment, packaging, transport or storage of such food results, or may be reasonably expected to result, in it or its by-products becoming directly or indirectly a component of such foods" (2).

Known as the additives "framework" Directive, this Directive also defines processing aids as "any substance not consumed as a food ingredient by itself, intentionally used in the processing of raw materials, foods or their ingredients, to fulfil a certain technological purpose during treatment or processing, and which may result in the unintentional but technically unavoidable presence of residues of the substance or its derivatives in the final product, provided that these residues do not present any health risk and do not have any technological effect on the finished product."

Processing aids

Whilst many of the substances used as additives may also be used as processing aids, the latter function is outside the scope of additives legislation. The differentiating criterion, and the question that any manufacturer must ask in terms of regulatory requirements, is "does it continue to function in the final food?" So, for example, sulphur dioxide (E220) may be used to prevent discoloration of fruit destined for pie making, but would have no effect in the fruit pie itself, and indeed would be cooked off during processing. Thus, in this application, it is a processing aid used in the making of a fruit pie, not an additive performing a function in the pie itself. Many of us will be used to similar techniques in the kitchen, such as using lemon juice to prevent discoloration. In the complex world of food manufacture, where production is increasingly specialised and expertise focused at specific sites, it is not unusual for the manufacturer of an end product to buy in many of his supplies as part-processed proprietary ingredients. So additives may be needed at the "intermediate" stage, but would have no function in the final product, and would therefore not appear on the label, unless considered to have the potential to cause an allergenic reaction (see Chapter 2). Thus, anti-caking agents may be required in dry ingredients to prevent them from turning lumpy before being made into a fancy cake, but will have no effect once the cake is baked and decorated, so the anti-caking agent functions as an additive in the dry mix, but is a processing aid as far as the cake is concerned. Other examples of processing aids are release agents used to prevent food from sticking to a mould or, perhaps, slicing equipment. Again, this is part of the process of production, not the composition of the food, even though there may be traces of the "processing aid" left on the product, as there would be on a cake from greasing the cake tin. This, then, is the essential technical difference between a processing aid and an additive.

The "framework" Directive identifies a number of classes of additives, e.g. sweeteners, colours and "miscellaneous" additives (including additive categories such as preservatives, antioxidants, emulsifiers, stabilisers, thickeners, flavour enhancers etc.), for which more detailed legislation was eventually developed, and lays down general criteria for their use, notably that technological need must be demonstrated that cannot be achieved by other means; that their presence presents no hazard to the consumer; and that they do not mislead the consumer. Their use may be considered only where there is demonstrable benefit to the consumer, namely to preserve the nutritional quality of the food; to provide necessary ingredients or constituents for foods manufactured for groups of consumers with special dietary needs, or to enhance the keeping quality or stability of a food or to improve its organoleptic properties, provided that, in doing so, it does not deceive the consumer; and to assist in manufacture, processing, preparation, treatment, packing, transport or storage of food, provided that the additive is not used to disguise the effects of the use of faulty raw materials or of undesirable (including unhygienic) practices or techniques during the course of any of these activities. These are similar to the principles enshrined in the Codex Alimentarius, the joint FAO/WHO body responsible for international standards in food.

The harmonisation of European legislation was a prerequisite for trade in the Single Market as differences in national legislation constituted barriers to trade. This is explored in greater detail in a later chapter, but it is important to appreciate that the development of a new raft of additives legislation in the late 1980s and through the 1990s was not indicative of an absence of controls before that time, but a recognition that differences in national approaches throughout the Member States were not conducive to the free movement of goods within a single economic entity. The new legislation reinforced the requirement for justification of a case of need in the use of additives and of the importance of not deceiving the consumer.

The primary aim of the food-manufacturing industry is to provide a wide range of safe, wholesome, nutritious and attractive products at affordable prices all year round in order to meet consumer requirements for quality, convenience and variety. It would be impossible to do this without the use of food additives. They are essential in the battery of tools used by the food manufacturer to convert agricultural raw materials into products that are safe, stable, of consistent quality and readily prepared and consumed.

Different types of additive are used for different purposes, though many individual additives perform more than one function. For the purposes of both classification and regulation, they are grouped according to their primary function. The main groupings, or classes, of additives are explained below, together with their functions and some examples of their use.

Preservatives

Preservatives are probably the single most important class of additives, as they play an important role in the safety of the food supply. Despite this fact, any chemical used to counteract the perishability of food raw materials has often become perceived as suspect, and any food containing a preservative has been considered inferior or unsafe. Yet the use of chemical preservatives, such as sulphur dioxide and sulphites, is but a continuation of the age-old practices of using salt, sulphite and spices to preserve perishable foods in the days before refrigeration and modern processing techniques. All food raw materials are subject to biochemical processes and microbiological action, which limit their keeping qualities. Preservatives are used to extend the shelf-life of certain products and ensure their safety through that extended period. Most importantly, they retard bacterial degradation, which can lead to the production of toxins and cause food poisoning. Thus they offer a clear consumer benefit in keeping food safe over the shelf-life of the product, which itself may be extended by their use and thus meet the demands of modern lifestyles, including infrequent bulk shopping expeditions. The continued perception of preservatives as undesirable, to which the many labels protesting "no artificial preservatives" testify, is therefore an unfortunate consumer misapprehension.

Antioxidants

Antioxidants reduce the oxidative deterioration that leads to rancidity, loss of flavour, colour and nutritive value of foodstuffs. Fats, oils, flavouring substances, vitamins and colours can all oxidise spontaneously with oxygen when exposed to air. The rate of deterioration can vary considerably and is influenced by the presence of natural antioxidants and other components, availability of oxygen, and sensitivity of the substance to oxidation, temperature and light, for example. Oxidation can be avoided, or retarded, by a number of means, such as replacing air by inert packaging gases, removal of oxygen with glucose oxidase, incorporation of UV-absorbing substances in transparent packaging materials, cooling, and use of sequestering agents. These may not be possible in all cases, or sufficient for an adequate shelf-life for some foods. Thus antioxidants are used to retard oxidative deterioration and extend shelf-life. Some antioxidants actually remove oxygen by self-oxidation, e.g. ascorbic acid, whilst others interfere in the mechanism of oxidation, e.g. tocopherols, gallic acid esters, BHA and BHT. All have specific properties, making them more effective in some applications than in others. Often a combination of two or more antioxidants is more effective than any one used simply because of their synergistic effects. The presence of sequestering agents, such as citric acid, may also have a synergistic effect, by reducing the availability of metallic ions that may catalyse oxidation reactions. The use of the powerful synthetic antioxidants BHA, BHT and the gallic acid esters is very restricted. Tocopherols, which can be either natural or synthetic, are less restricted but are less effective in the protection of processed foods. Antioxidants cannot restore oxidised food; they can only retard the oxidation process. As oxidation is a chain reaction process, it needs to be retarded as early as possible. The most effective use of antioxidants is therefore in the fats and oils used in the manufacturing process.

Emulsifiers and stabilisers

The purpose of emulsifiers and stabilisers is to facilitate the mixing together of ingredients that normally would not mix, namely fat and water. This mixing of the aqueous and lipid phases is then maintained by stabilisers. These additives are essential in the production of mayonnaise, chocolate products and fat spreads, for example. The manufacture of fat spreads (reduced-fat substitutes for butter and margarine), has made a significant contribution to consumer choice and dietary change, and would not be possible without the use of emulsifiers and stabilisers. Other reduced- and low-fat versions of a number of products are similarly dependent on this technology. Anyone who has ever made an emulsified sauce, such as mayonnaise or hollandaise, will appreciate the benefits of this technology – still more so those who have failed miserably in the technique and ended up with an expensive mess of curdled ingredients!

In addition to this function, the term stabiliser is also used for substances that can stabilise, retain or intensify an existing colour of a foodstuff and substances that increase the binding capacity of the food to allow the binding of food pieces into reconstituted food.

The increasing awareness of problems with food allergy and intolerance has led to the requirement to state the source of certain emulsifiers on food labelling. For example, lecithin derived from soya is not suitable for an individual with an allergy to soya, therefore clear labelling of the source of the ingredient is vital to aid in consumer choice of products safe for individuals with specific dietary requirements (see Chapter 2).

Colours

Colours are used to enhance the visual properties of foods. Their use is particularly controversial, partly because colour is perceived by some as a means of deceiving the consumer about the nature of the food, but also because some of the most brightly coloured products are those aimed at children. As with all additives, their use is strictly controlled and permitted only where a case of need is proven, e.g. to restore colour that is lost in processing, such as in canning or heat treatment; to ensure consistency of colour; and for visual decoration. The use of colour in food has a long and noble tradition in the UK. Medieval cooks were particularly fond of it. The brilliant yellow of saffron (from which Saffron Walden derives its name) and the reddish hue of saunders (powdered sandalwood) were used along with green spinach and parsley juice to colour soups in stripes or to give marbleised effects (1). So, whilst adding colour to food may appear to some to be an unnecessary cosmetic, which is not in the consumer's interests, there can be no doubt that the judicious use of colour enhances the attractiveness of many foods. Some retailers tried introducing ranges of canned vegetables and fruits such as strawberries and peas without adding back the colour leached out by heat processing. They were still trying to dispose of the unsold returns several years later! Colour is important in consumer perception of food and often denotes a specific flavour. Thus, strawberry flavour is expected to be red and orange flavour orange-coloured. Consumer expectation is therefore a legitimate reason for adding colour.

Food colourings, in particular, have long been the scapegoat in the popular press for behaviour problems in children. It has been over 30 years since Feingold suggested that artificial food colours and preservatives had a detrimental effect on the behaviour of children (3).

Since then, research into the effect of colours and preservatives in foods on children's behaviour has added fuel to the fire of negative consumer perception of these additives, particularly in products aimed specifically at this age group (4). Significant changes were found in the hyperactivity behaviour of children by removing colorants and preservatives from the diet. There was no gender difference in this result and the reduction of hyperactivity was independent of whether the child was initially extremely hyperactive, or not hyperactive at all. More recently in 2007, a study on the effect of two mixtures of certain artificial food colours together with the preservative sodium benzoate showed an adverse effect on the hyperactive behaviour of children in some age groups in comparison with a placebo, although the increases in the levels of children's hyperactive behaviour were not consistently significant for the 2 mixtures or in the 2 age groups (5). The findings of this new study replicate and extend the findings from an earlier study in preschool children in 2004 (6). The colours used in this study are already included in work of the European Food Safety Authority (EFSA) on the re-evaluation of colours.

Colouring Foodstuffs

The term 'colouring foodstuffs' has been adopted for colourings that are derived from recognised foods and processed in such a way that the essential characteristics of the food from which they have been derived are maintained.

This is a different situation to natural colours that are regarded as additives where the pigment is selectively extracted and concentrated.

A colouring foodstuff can be declared as an ingredient on the label without a requirement for its function to be listed, as legislation only requires this of additives.

These colouring foodstuffs include bright yellow colours derived from turmeric, oleoresin and safflower; golden yellow to natural orange colours from carrots and paprika; toffee brown colour from caramelised sugar syrup; green colours from spinach leaves and stinging nettles, both rich in chlorophylls; and red, blue and purple colour from concentrates of red and blue fruits, red cabbage and beetroot, rich in anthocyanins.

It is clear that the full spectrum of colour shades is achievable using colouring foodstuffs, although developers should ensure that the colouring foodstuff exhibits the same stability and vibrancy of colour in the final application as a conventional food colouring would.

Sweeteners

Sweeteners perform an obvious function. They come in two basic types – "bulk" and "intense", and are permitted in foods that are either energy-reduced or have no added sugar. They are also sold direct to consumers as "table-top" sweeteners – well-known to dieters and diabetics. For example the table top sweetener Sunette contains acesulfame-K while Splenda contains sucralose. Intense sweeteners, such as aspartame, saccharin, acesulfame-K and sucralose have, as their name suggests, a very high sweetening property, variable from type to type but generally several magnitudes greater than that of sucrose. (For example, aspartame is approximately 200 times sweeter than sugar, weight for weight; saccharin 300–500 times; and acesulfame-K 130–200 times.) Bulk sweeteners, where the majority are polyols, including erythritol, sorbitol, isomalt and lactitol are less sweet, but provide volume and hence mouthfeel. Amongst the polyols, maltitol is one of the sweetest and xylitol, which is the sweetest, has the same sweetness intensity as sucrose. Due to the reduced sweetness characteristics of the majority of polyols, it is possible to blend them with other polyols or with intense sweeteners to improve the sweetness and taste quality. This property is known as sweetness synergy. Another benefit is the ability to mask the undesired bitter metallic aftertaste of some intense sweeteners. Commonly used combinations include, saccharin with cyclamate, acesulfame-K with aspartame, erythritol with acesulfame-K and there are many more. Both types of sweetener (bulk or intense) are useful in low-calorie products, and are increasingly sought after by many consumers, and for special dietary products such as for diabetics. The absence of sucrose also lowers the cariogenic properties of the product.

Flavour enhancers

This is a group of additives that has attracted adverse attention, in particular monosodium glutamate (MSG:E621), which is widely blamed for an intolerance reaction that became known as "Chinese Restaurant Syndrome".

Flavour enhancers are substances that have no pronounced flavour or taste of their own but which bring out and improve the flavours in the foods to which they are added. Although salt has a distinctive taste of its own and is not classed as a food additive, it is in fact the most widely used flavour enhancer. The next best known is glutamic acid and its salts, most commonly found in the form of monosodium glutamate, which has been used for several centuries in the Far East as a condiment in savoury products. It is a normal constituent of all proteins, an essential amino acid and present in the body. The alleged intolerance reaction was never confirmed in sound scientific studies. Anyone showing a reaction to MSG used as an additive would necessarily also react to foods that contain it naturally in high quantities, such as tomatoes and cheese.

Some sweeteners have also been found to have flavour-enhancing properties and have been authorised for use as such. For example, neohesperidine DC (E959) can enhance the flavour of meat products and margarine, and acesulfame K, aspartame and thaumatin are used to enhance the flavour of chewing gum and desserts.

Flavourings

Although flavour enhancers are categorised as additives, flavourings are technologically different and regulated separately, even though they are often considered by the general public to be the same thing. Flavourings are defined as imparting odour and/or taste to foods and are generally used in the form of mixtures of a number of flavouring preparations and defined chemical substances. These do not include edible substances and products intended to be consumed as such, or substances that have exclusively a sweet, sour or salty taste, i.e. ordinary food ingredients such as sugar, lemon juice, vinegar or salt. The latest draft of the proposed new EC Regulation on Flavourings would also exclude from the definition of flavourings raw foods and non-compound foods, and mixtures of spices or herbs, mixtures of tea provided they are not used as food ingredients. In addition to the types of flavouring such as process flavours or smoke flavours, there are three distinct classes of flavouring substances: natural, e.g. citral; nature-identical, e.g. vanillin; and artificial, e.g. ethyl vanillin. Some 2700 substances were identified and included in a European register following Commission Decision (EC) 1999/217/EC as amended. Then there are flavouring preparations, e.g. vanilla extract. Many flavourings are sold as a complex mixture of individual preparations and flavouring substances, generally confidential to the company that has produced the flavouring. Legislation has been designed to protect commercial confidentiality in registering on the EC list newly discovered flavouring substances. Because of the complexity of the flavouring used in a food, labels generally indicate simply "flavourings" in the ingredients list. This is all that is legally required, as to list every individual substance would often be extremely lengthy and virtually incomprehensible to the consumer, although the manufacturer may be more specific if he wishes. Any flavourings labelled as "natural" must meet the legal definition. The Food Standards Agency has issued criteria for the use of the term "natural" in product labelling. The new proposal for an EC Regulation on flavourings and certain food ingredients with flavouring properties for use in and on foods means that in future there are likely to be stricter controls for the labelling of natural flavourings (7).

As with additives, some flavourings are sold direct to the consumer for domestic culinary use. Vanilla and peppermint are amongst the best known, as well as the popular brandy and rum essences. Anyone who has ever added too much flavouring to a home-made cake or a batch of peppermint creams will appreciate the minute quantities in which they are used. Similarly, in commercial manufacture, the quantity of flavouring used is extremely small in relation to that of other ingredients. Most flavourings are developed from substances naturally present in foods. Citrus and orange oils, for example, are amongst the most common natural source materials used in flavouring preparations and substances.

Other additives

Colours and sweeteners are very specific, well-defined classes of additives and, because of the nature of their function, are subject to specific legislation. All other classes of additive now fall under the general heading of "miscellaneous". In addition to the larger groups mentioned above, there are other categories within this more general grouping – namely thickeners, acids, acidity regulators, anti-caking agents, anti-foaming agents, bulking agents, carriers, glazing agents, humectants, raising agents and sequestrants.

The function of most of these is obvious from the name, with the possible exception of sequestrants. These are substances that form chemical complexes with metallic ions. They are not widely used and this is a class of additives rarely seen on a food label. Thickeners, on the other hand, are amongst the most commonly used additives, as they exert an effect on the texture and viscosity of food and drinks products. Much as various types of flour are used extensively in the kitchen to thicken sauces, soups, stews and other dishes with a high liquid content, most commercial thickeners are starch- or gum-based and serve much the same purpose.

One class of additive that has no domestic equivalent is that of packaging gases. These are the natural atmospheric gases now widely used in certain types of pre-packed products, such as meat, fish and seafood, fresh pastas and ready-prepared vegetables found on the chilled food counters in sealed containers. The "headspace" of the container is filled with one or a combination of the gases, depending on the product, to replace the air and modify the atmosphere within the pack to help retard bacteriological deterioration, which would occur under normal atmospheric conditions – hence the term "packaged in a protective atmosphere". Arguably, the gases do not have an additive function as they are not detectable in the food itself and function only to preserve the food for longer in its packaged state, but for regulatory purposes they were deemed to be additives and must therefore be labelled. Carbon dioxide will, of course, also be familiar as an ingredient in many fizzy drinks - an illustration of the many different functions and uses of additives.

Current EC legislation on additives does not cover the use of enzymes apart from invertase and lysozyme. However, in July 2006, the European Commission published a package of legislative proposals to introduce harmonised EU legislation on food enzymes for the first time and upgrade current rules for food flavourings and additives to bring them into line with the latest scientific and technological developments. The proposals were amended in October 2007 and are discussed further in the next chapter.

Safety of Additives

The safety of all food additives, whether of natural origin or synthetically produced, is rigorously tested and periodically re-assessed. In the UK, the responsible authority is the Committee on Toxicity of Chemicals in Food, Consumer Products and the Environment (COT), a Government-appointed expert advisory committee, which provides advice to the Food Standards Agency, the Department of Health and other Government Departments and Agencies on matters concerning the toxicity of chemicals, including food additives. At

European level, all additives approved for use in legislation have been evaluated by the Scientific Committee on Food (SCF) or, since May 2003, by its replacement the European Food Safety Authority (EFSA). Therefore, EFSA's Panel on Food Additives, Flavourings, Processing Aids and Materials in Contact with Food (AFC panel) is now responsible for the safety evaluation of new food additives (8,9).

Only additives evaluated in this way are given an "E" number; thus the "E" number is an indication of European safety approval, as well as a short code for the name of the additive.

In evaluating an additive, EFSA allocates an "Acceptable Daily Intake" (ADI), the amount of the substance that the panel considers may be safely consumed, daily, throughout a lifetime. This assessment is used to set the maximum amount of a particular additive (or chemically related group of additives) permitted in a specific food, either as a specified number of grams or milligrams per kilogram or litre of the food or, if the ADI is very high or "non-specified", at *quantum satis*, i.e. as much as is needed to achieve the required technological effect, according to Good Manufacturing Practice.

In establishing the ADI, a safety factor is always built in, usually 100-fold, to ensure that intake of any additive is unlikely to exceed an amount that is anywhere near toxicologically harmful. To ensure that consumers are not exceeding the ADI by consuming too much of or too many products containing a particular additive, the EU legislation requires that intake studies be carried out to assess any changes in consumption patterns.

The UK has carried out a number of intake surveys involving specific additives. None has culminated in results that have given cause for concern, except that in its 1994 survey of artificial sweeteners, consumption by some toddlers was considered to be excessive, given their high consumption of fruit squash. This potential problem was resolved by advice to add extra water to squash given to toddlers. It also raised questions about the establishment and application of the ADI, given that it is intended to cover changes in patterns of eating throughout a lifetime, from weaning to old age, but that is a separate scientific debate in itself.

At international level, there is a further level of evaluation of food additives, contaminants and residues of veterinary drugs in food by the Joint Expert Committee on Food Additives (JECFA), which advises the UN's Food and Agriculture Organization (FAO) and World Health Organization (WHO) Codex Alimentarius, which sets international standards. This has become increasingly important in recent years as World Trade Organisation (WTO) arrangements specify that Codex standards will apply in any dispute over sanitary and phytosanitary standards, i.e. the safety and composition of foods. For this reason, the Codex General Standard for Food Additives (GSFA), was adopted to recommend usage levels of food additives in all products traded internationally.

As part of EFSA's role in the area of food additives, it is involved in the re-evaluation of all authorised food additives in the EU.

In September 2004, EFSA issued an opinion on the safety of parabens (E214-219) used as preservatives in foods following a risk assessment of its use in foods. As a result, Directive 2006/52/EC amending Directive 95/2/EC on food additives other than colours and sweeteners and Directive 94/35/EC on sweeteners for use in foodstuffs, deleted the preservatives, E216 propyl p-hydroxybenzoate and E217 sodium propyl p-hydroxybenzoate from the list of permitted preservatives in Annex III (10).

In the area of sweeteners, the safety of aspartame was considered controversial, especially following a long-term study on its carcinogenicity in 2005. Hence, EFSA evaluated findings from this study, and, in this case, confirmed that there was no need to revise the previously established ADI (11).

On the other hand, in re-evaluating the colour E128, Red 2G, in 2007, EFSA decided that there was a safety concern, and later the Commission suspended its use (12,13).

Intolerance

Additives have often been blamed for causing intolerance or allergic reactions, especially hyperactivity in children. Whilst there is no doubt that certain foods and food ingredients, including additives, are responsible for intolerance reactions, the prevalence of such reactions has often been greatly exaggerated. Genuine intolerance to food additives is extremely rare. It has been estimated that the true prevalence of intolerance to foods is about 2% in adults and up to 20% in children, and for food additives from 0.01 to 0.23%. The substantial overestimation of such reactions by the general public probably owes itself to the adverse media coverage and anti-additives campaigning of the 1980s, when popular belief was that additives were responsible for harmful behavioural effects and hyperactivity was attributed solely to the consumption of tartrazine (E102). The result was that tartrazine, an azo (synthetic) colour, was removed from a wide range of products, especially sweets and soft drinks that were likely to be consumed by children, as consumers in their droves ceased to buy anything that was labelled as containing it. Manufacturers are still reluctant to use this colour, unless there is nothing else in the palette of yellow colours authorised for the product. Such is the power of consumer choice, be it informed or otherwise.

Food intolerance, and especially allergy, is again under the spotlight, not now because of alleged hyperactivity in children, but, far more seriously, because of the seemingly growing prevalence of severe allergic reactions, particularly to peanuts. Since the mid-1990s, there have been a number of widely reported incidents, including several tragic deaths as a result of anaphylactic shock, a severe allergic reaction to specific proteins, most commonly those found in tree nuts and peanuts and a small number of other foods, including milk, wheat, eggs, soya, fish and shellfish. The reasons for such reactions are not yet fully understood and are still under investigation, as are the causes of this apparently growing problem, but the need to address the issue and do everything possible to assist the small but significant number of people affected by this most severe form of allergy caused the European Commission to task its former Scientific Committee for Food (SCF) with identifying the scope of the problem and the foods and ingredients associated with it. This 1996 Report reaffirmed the SCF's earlier (1982) estimation of intolerance to additives as affecting from 0.01 to 0.02% of the European population (14). More specifically, the prevalence of intolerance to food additives in the population was put at 0.026%, or about 3 people per 10,000 of the population. This compares with the prevalence of adverse reactions to cows' milk of 1 to 3%. The most commonly observed reaction is now to sulphur dioxide (E220) and sulphites, especially in asthma sufferers, again growing in number or perhaps being more frequently reported.

It must be understood that the incidence of genuine intolerance to additives is very low. Accurate labelling is the key to avoiding unnecessary suffering of an adverse reaction, such as urticaria, asthma or atopic symptoms, in the case of sensitised consumers, or adverse publicity in the case of food producers, and for this reason the EC Labelling Directive 2000/13/EC was amended in 2002, 2003, 2006 (to establish a list of potential allergens that must be declared by name on food labels) and in 2007 (see Chapter 2).

Myths and Fallacies

Nothing is guaranteed to fill column inches and dominate the airwaves more than a good food scare. Additives have seen their share of these, though not on the scale of the 1996 BSE crisis or the more recent controversy over genetic modification: though equally long-running and bearing similarities to the latter issue, additives were never the butt of a concerted campaign by environmentalists and others dedicated to the downfall of a specific technology. Anti-additives campaigns would either target a specific additive or class of additives, for whatever reason, or cite the use of additives as part of a general thrust to disparage the modern food-manufacturing industry and seek to encourage a "back to basics" trend towards good old-fashioned home cooking and away from the purported less healthy foods produced by industrial processing for the UK's largely urbanised society.

Hence the periodic targeting of preservatives, antioxidants, azo colours, sweeteners and monosodium glutamate. The evidence of such "scares" still abounds on the labels of countless products that claim to be free from "artificial" preservatives, colours and additives in general. This is indicative of the susceptibility of both marketing men and consumers to perceived adverse effects of particular additives. Such a response is unhelpful; whilst it is understandable that consumer concern in response to a media scare may result in a company removing an additive, or indeed any other ingredient, from a product for reasons of short-term expediency, the options and alternatives will inevitably become reduced every time something is removed from the range of ingredients, and the controversy left unresolved. It would be far better to address the issue through appropriate scientific investigation and seek to ensure that evidence of safety and absence of adverse effects are given at least some airing in the public domain to explode the myth engendered by the original controversy.

This, of course, is not easy, as good news is, generally speaking, no news at all and certainly unlikely to make the headlines. The tabloid newspapers had a field day with the Food Commission's stories that "Cyclamates 'may cause testicular atrophy'" (15) and "Aspartame 'may cause brain tumours'" (16). Refuting such headlines is not easy; the full barrage of scientific evidence generally needs to be brought out in defence of any food ingredient or additive placed under the media spotlight and accused of causing some adverse effect. Often the "evidence" produced in support of the story needs to be pulled apart under the microscope and any deficiencies, such as in the research protocols or the way in which any experimentation was conducted, identified. The motivation for publishing such "research", and any exaggeration of the findings, also need to be examined.

All this takes time and will not protect any company using the additive or additives concerned from a barrage of enquiries from worried customers who, not unnaturally, seek reassurances that they have not already been harmed or will not be if they continue to consume the product. Again, a sense of proportion is important. The "problem" needs to be placed in context, given perspective against the wide range of risk factors to which all of us are exposed in daily life, and consumers assisted and encouraged to develop their own sense of risk assessment and risk management. This will become all the more important as communication becomes ever more global and instantaneous. The internet offers both threats and benefits: threats in that anyone can rapidly set off a scare by posting adverse information about, say, a specific sweetener. This may be a genuine concern that some possible risk to, perhaps, a certain sector of the population has been found, maybe to people suffering from a specific condition. It may also be that an unscrupulous company seeking to target that group with a new product decides to set off a scare shortly before launching its product, which is marketed as "free from" that additive or ingredient. The benefit lies in being able to expose such scares equally quickly, and the opportunity to post true and accurate information about food production for those who want to know.

Clean Labels

The growing demand from health-conscious consumers is for the replacement of artificial food additives with 'natural' ingredients, which perform similar technological functions. Thus, food processors are continuously seeking natural alternatives to food additives as, when these are listed on labels as the named ingredients rather by E-number, it gives the food product a 'clean label' declaration.

Clean label declarations are not regulated; however, the Food Standards Agency in the UK has issued "Criteria for the use of the terms Fresh, Pure, Natural etc." which could be used as guidance. In addition, when incorporating new substances into foods one would also need to comply with the EC Regulation 258/97 concerning Novel Foods and Novel Food Ingredients.

A number of ingredients are now being manufactured that claim to give foods a clean label status e.g. emulsifiers such as lecithin and soya protein; antioxidants including grape seed, chestnut and olive leaf extracts; colours for example, lycopene, anthocyanin and chlorophyll; and preservatives including cinnamic acid, carvacol, chitosan, and lysozyme.

Some bacterial cultures, known as 'protective cultures', able to inhibit the growth of pathogenic bacteria and mycotoxin-producing mould are being used as inhibitors of foodborne microorganisms. These protective cultures produce antimicrobial metabolites like organic acids (lactic and acetic acid), and bacteriocins (nisin and natamycin), and are substitutes for conventional additives, helping manufacturers make the 'Clean Label' claim.

It will be some time before we see a complete shift to clean label products, and in some situations this may not be possible due to a lack of suitable natural alternatives.

Conclusions

Much has happened to and in the food industry and the market for food since the great focus on additives in the 1980s. The popular books produced on the subject at that time focused largely on the potential adverse effects of additives; the potential misleading of consumers about the food they were eating; and the profit-driven nature of the industry motivated to use additives in their products (17,18,19). But not all of this criticism was without justification, and there were undoubtedly bad practices in place in some sectors of the industry, where unscrupulous traders saw opportunities for quick profit. The use of phosphates in reconstituted meat and fish products to make them appear as better-quality cuts and fillets or to add weight to a chicken was a dodge that trading standards officers rightly pursued with some zeal. This is not a criticism of the legitimate use of phosphates in meat products such as hams, but of the instances of false description of reconstituted products as prime cuts, and frozen "scampi" that disintegrated on defrosting. Any business will always have its unscrupulous operators, but strict regulation and enforcement now make this increasingly difficult in the food industry.

The 1990 Food Safety Act provided the framework of primary legislation for the food industry in the UK. The raft of legislation on food additives developed as part of the European Single Market, and explored in detail in a later chapter, strictly controls the use of all additives.

The establishment of the Food Standards Agency, with its dual role of protecting and informing the consumer, may well influence both trends in the use of additives and public perceptions of their worth.

Furthermore, the market has changed considerably in recent years, partly as a result of European integration and partly because consumers have become more sophisticated, more knowledgeable, and more affluent. Overseas travel has greatly broadened the British palate and increased demand for a wide range of exotic and adventurous foods that have been sampled overseas. Our increasingly cosmopolitan society has also led to the availability of more and more "ethnic" foods, both in restaurants and for domestic consumption, while busy lifestyles, and the increasing number of working women have led to more and more food being consumed outside the home.

Never has the range and choice of foods been so great, in terms of availability in the supermarkets and specialist food shops, or through the catering trade. This is not to say that additives are less widely used or less relevant – far from it. But those who wish to avoid them, either as manufacturers or consumers, should find it possible to do so, and those who do use them need have no concerns, except to obey the law in the case of manufacturers, and to understand the meaning of the ingredients list in the case of consumers. Astute consumers now notice that it is not only pre-packed foods that contain additives: foods sold "loose" at delicatessen counters are now also labelled to indicate the content of additives – or should be. And it has not escaped the notice of public health analysts that the greatest use of food colours is in ethnic restaurants. Public protection is ensured and additives cannot be used to deceive, but we would be deceiving ourselves if we thought that we could continue to enjoy the choice, ease and convenience of our food supply without them. Like them or not, they are a fact of life and their usefulness cannot be denied.

References

1   McKendry M. Seven Hundred Years of English Cooking. London, Treasure Press. 1973.

2   Directive 89/107/EEC on the approximation of the laws of the Member States concerning food additives authorised for use in foodstuffs intended for human consumption, as amended. The Official Journal of the European Communities. 1989, 32 (L40), 27-33.

3   Feingold B.F. Hyperkinesis and learning disabilities linked to artificial food flavors and colors. *American Journal of Nursing*, 1975, 75, 797–803.

4   Dean T. Do food additives cause hyperactivity? in Food Allergy and Intolerance, Current Issues and Concerns. Ed. Emerton V. Leatherhead. Leatherhead Food International, 2002, 93-101.

5   McCann D *et al.* Food additives and hyperactive behaviour in 3-year-old and 8/9-year-old children in the community: a randomised, double-blinded, placebo-controlled trial. *The Lancet*, 2007, 370, 1560-7.

6   Bateman B *et al.* The effects of a double blind, placebo controlled, artificial food colourings and benzoate preservative challenge on hyperactivity in a general population sample of preschool children. *Archives of Disease in Childhood*, 2004, 89, 506-11.

7   Proposal for a Regulation of the European Parliament and of the council on flavourings and certain food ingredients with flavouring properties for use in and on foods (amended proposal - 24 October 2007) http://eur-lex.europa.eu/LexUriServ/site/en/com/2007/com2007_0671 en01.pdf

8   Minutes of the 2nd Plenary meeting of the Scientific Panel on food additives, flavourings, processing aids and materials in contact with food - Held in Brussels on 9 July 2003 http://www.efsa.europa.eu/EFSA/Event_Meeting/minutes_afc_02_ adopted_en1,0.pdf

9   Report from the Commission to the European Parliament and the Council on the progress of the re-evaluation of Food Additives (July 2007) http://eur-lex.europa.eu/LexUriServ/site/en/com/2007/com2007_0418en01.pdf

10   Opinion of the Scientific Panel on food additives, flavourings, processing aids and materials in contact with food (AFC) related to para hydroxybenzoates (E 214-219) (Adopted on 13 July 2004)

11   Opinion of the Scientific Panel on Food Additives, Flavourings, Processing Aids and Materials in contact with Food (AFC) on a request from the Commission related to a new long-term carcinogenicity study on aspartame (Adopted on 3 May 2006)

12   Opinion of the Scientific Panel on Food Additives, Flavourings, Processing Aids and Materials in Contact with Food on the food colour Red 2G (E128) based on a request from the Commission related to the re-evaluation of all permitted food additives (Adopted on 5 July 2007)

13   Commission Regulation (EC) No 884/2007 of 26 July 2007 on emergency measures suspending the use of E128 Red 2G as food colour

14   Scientific Committee for Food. Report on Adverse Reactions to Food and Food Ingredients. 1996.

15   Anon. Cyclamate levels 'may cause testicular atrophy'. The Food Magazine. 1997, 36, 1.

16   Anon. Aspartame 'may cause brain tumours'. The Food Magazine. 1997, 36, 5.

17   Millstone E. Food Additives – Taking the lid off what we really eat. Harmondsworth, Penguin. 1986.

18   Hanssen M. E for Additives – The complete E number guide. Wellingborough, Thorsons. 1984.

19   Saunders B. Understanding Additives. London, Consumers Association. 1988

Further Reading

Wilson R. Ingredients Handbook – Sweeteners (3rd Edition). Leatherhead Food International, 2007.

Kendrick A. Clean up your label with the colourful alternatives to additives, *Confectionery Production*, 2005, 71(4), 14-15.

Anon. Colouring Foodstuffs – The Clean Label Colourful Alternative to Additives? *Innovations in Food Technology*, 2006, 30, 76-7.

Stich E., Court J., Colouring without Colour, *Fruit Processing*, 2006, 16 (3), 161-5.

The European Food Safety Authority, Food Additives, 2007
http://www.efsa.europa.eu/EFSA/KeyTopics/efsa_locale-1178620753812_FoodAdditives.htm..

Emerton V. Ingredients Handbook – Food Colours (2nd Edition). Leatherhead Food International, 2008.

2. WHAT SHOULD BE DECLARED ON THE LABEL

Introduction

The primary legislation that needs to be examined when looking at labelling of food additives is the Food Safety Act 1990, which stipulates that it is illegal to sell food that is 'injurious to health' or to falsely describe it. Similar provisions are set in the General Food Regulations 2004. Other legislation that also needs to be considered is the Trade Descriptions Act 1968, the Weights and Measures Act 1985, and the Food (Lot Marking) Regulations 1996. The main regulations that relate specifically to labelling of food additives are the Food Labelling Regulations 1996 (as amended) and the Food Additives Labelling Regulations 1992.

This chapter gives a basic outline of what should be on the label of a pre-packed food, as specified in the Food Labelling Regulations. Details are given on what should be declared, including the name of the food, the list of ingredients, the appropriate durability indication, a quantitative ingredients declaration, storage conditions, place of origin and instructions for use. The requirements of the Food Additives Labelling Regulations are outlined and relevant areas that define food additives and prescribe requirements for labelling of food additives for business and consumer sale are highlighted. Please note that the actual legislation should be consulted when constructing or checking label copy.

Food Safety Act 1990

It is worth noting that, although the Food Safety Act does not contain details of labelling requirements, it does set an overarching provision prohibiting labelling of food with a false description or a description that may mislead the consumer as to the nature, substance or quality of the food.

The Act makes it an offence for anyone to sell, or possess for sale, food that:

- has been rendered injurious to health;
- is unfit or so contaminated that it would be unreasonable to expect it to be eaten;
- is falsely described, advertised or presented;
- is not of the nature, substance or quality demanded.

The General Food Regulations 2004 (S.I. 2004 No. 3279)

These regulations enforce certain provisions of Regulation (EC) No. 178/2002 laying down the general principles and requirements of food law.

* Article 14 specifies food safety requirements and prohibits the placing of unsafe food on the market
* Article 16 states that the labelling/advertising/presentation of food should not mislead consumers
* Article 18 states that the traceability of food/any other substance to be incorporated into a food shall be established at all stages of production, processing and distribution and sets obligations to food business operators.
* Article 19 places an obligation on food business operators to take responsibility and initiate the withdrawal of food if it does not comply with food safety requirements.

Trade Descriptions Act 1968

The Trade Descriptions Act 1968 makes it an offence for a trader to

* apply a false trade description to any goods;
* supply or offer to supply any goods to which a false trade description is applied.

Parts of the Trade Descriptions Act will soon be amended or repealed by virtue of the UK implementation of the Unfair Commercial Practices Directive 2005/29/EC. This Directive harmonises unfair trading laws in all EU member states and its provisions must have been applied in member states by 12 December 2007.

In the UK, it is expected that the regulation implementing this Directive, the Consumer Protection from Unfair Trading Regulations 2007, will come into force by April 2008.

Weights and Measures Act 1985

This Act provides for regulations to be drawn up on the expression of net quantity on prepacked food. The Act also provides for the 'average' system of quantity control for prepacked goods. Most foods and additives prepacked in quantities greater than 5 g or 5 ml need a quantity mark.

Food (Lot Marking) Regulations 1996 (S.I. 1996 No. 1502)

The aim of these Regulations is to establish a framework for a common batch or lot in order to facilitate the tracing and identification of product along the food chain.

Food Labelling Regulations 1996 (S.I. 1996 No. 1499 as amended by S.I. 1998 No. 1398, S.I. 1999 No. 747, S.I. 1999 No. 1483, S.I. 2003 No. 474, S.I. 2004 No. 1512, S.I. 2004 No. 2824, S.I. 2005 No. 899, S.I. 2005 No. 2057, S.I. 2005 No. 2969 and S.I. 2007 No. 3256.)

The general requirements laid down by the Food Labelling Regulations 1996 are set out below in more detail.

The Regulations stipulate that all prepacked foods that will be supplied to the ultimate consumer or to a catering establishment must be labelled with:

Name of the food

If there is a name prescribed by law for a food it must be used as the name of the food. If there is no name, a customary name may be used. If there is neither a name prescribed by law nor a customary name, the name must inform the buyer of the true nature of the food and, if necessary, must include a description of use. Trademarks or brand names may be used but these may not substitute for the name of the food. If necessary, the name must include an indication of the physical condition of the food or any treatment that it has undergone.

List of ingredients

All labels must include a list of ingredients, and all ingredients in the food must be declared in the list, unless there are specific exemptions. The title 'Ingredients' must be contained in the heading for the list.

The names given to these ingredients must be the same as if they were being sold as a food. If the ingredient has been irradiated in any way, its name must be accompanied by the word 'irradiated' or the phrase 'treated with ionising radiation'.

Ingredients must be listed in descending order of weight as used during the preparation of the food. The exception is water and volatiles, which should be listed in order of weight in the final product. Ingredients that are reconstituted during preparation may be included in the list of ingredients in the order of their weight before concentration or drying. However, if the food is a dehydrated or concentrated food, which will be reconstituted with water, then the ingredients may be listed as reconstituted.

If the food contains mixed fruit, vegetables, or mushrooms present in variable proportions but of similar weight, these ingredients may be grouped together in one place in the list of ingredients by their designation of 'fruit', 'vegetables' or 'mushrooms' and labelled with the words 'in variable proportions' followed by a list naming all the fruit, vegetables or mushrooms present.

If the food contains a mixture of herbs and spices and these are in equal proportion, these ingredients may be listed in any order and labelled with the words 'in variable proportions'.

Ingredients that constitute less than 2% of the finished product may be listed in a different order after the other ingredients. Similar or mutually substitutable ingredients and those used in the preparation of a food without altering its nature or its perceived value (excluding additives and allergenic ingredients) that make up less than 2% of the finished product may be referred to by the phrase 'contains...and/or...' where at least one of no more than two such ingredients is present in the finished product.

The Labelling Regulations include a list of permitted generic names that may be used to name an ingredient provided it meets the specified conditions, for example for oils the generic name 'vegetable oil' or 'animal oil' may be used rather than the specific source of the oil, provided an indication that the oil has been hydrogenated is given where appropriate.

If water constitutes more than 5% of the finished product, it must be included on the ingredients list.

For a list of foods that are exempt from ingredients listing, see Appendix A. It is important to note that the allergen labelling requirements detailed later override these exemptions.

Compound ingredients

If a compound ingredient (an ingredient composed of two or more ingredients, including additives) is used in the food, the names of the ingredients in the compound ingredient must be given in the list of ingredients either instead of the name of the compound ingredient or in addition to it.

If the name of the compound ingredient is given, its ingredients must immediately follow the name.

The names of the ingredients of a compound ingredient do not need to be listed if the compound ingredient:

- is a food that if sold by itself would not require a list of ingredients;
- is an ingredient which is identified by a permitted generic name;
- constitutes less than 2% of the finished product and its composition is defined in Community legislation (e.g. that on chocolate, fruit juice, jam, fat spreads); or
- constitutes less than 2% of the finished product and consists of a mixture of spices and/or herbs.
- If they are exempt when the compound food is sold as such.

It is important to note that the allergen labelling requirements detailed later override these exemptions.

If an ingredient of a compound ingredient has been irradiated, it must be listed and accompanied by the word 'irradiated' or words 'treated with ionising radiation' except in the case of food prepared for patients needing sterile diets under medical supervision.

Additives

Additives added to or used in a food to serve the function of one of the categories of additives listed below must be identified in the ingredients list by the name of the category followed by the specific name or serial number ('E number').

If an additive serves more than one function, it is only necessary to indicate the category that represents the principal function served by the additive in the food. If an additive serves none of these functions, it must be declared by its specific name in the ingredients list.

The following list shows the categories of additives that must be identified in a list of ingredients by their category name (Schedule 4 Food Labelling Regulations).

Acid [1]	Flour treatment agent
Acidity regulator	Gelling agent
Anti-caking agent	Glazing agent
Anti-foaming agent	Humectant
Antioxidant	Modified starch [2]
Bulking agent	Preservative

---

[1] In the case of an additive that is added to or used in food to serve the function of an acid and whose specific name includes the word 'acid', it is not necessary to use the category.

[2] Neither the specific name nor the serial number need be indicated. However, if the modified starch may contain gluten, the vegetable origin must be indicated, e.g. 'modified wheat starch'.

Colour

Emulsifier

Emulsifying Salts

Firming agent

Flavour enhancer

Propellant gas

Raising agent

Stabiliser

Sweetener

Thickener

Flavourings

If a flavouring is added to or used in a food, it should be described in the ingredients list using the word 'flavouring' or, where more than one flavouring ingredient is used, the word 'flavourings'. A more specific name or description of the flavouring may be used.

Use of the word 'natural'.

The word 'natural' or any word having substantially the same meaning, may be used for an ingredient being a flavouring only where the flavouring component of such an ingredient consists exclusively of:

- a flavouring substance (a defined chemical substance) that is obtained by physical (e.g. distillation and solvent extraction), enzymatic or microbiological processes, from material of vegetable or animal origin, which is either raw or subjected only to a normal process used to prepare food for human consumption; or
- a flavouring preparation, i.e. other products, possibly concentrates, obtained by physical, enzymatic or microbiological processes from material of vegetable or animal origin.

Processes normally used in preparing food for human consumption include drying and fermentation.

If the name of the flavouring refers to the vegetable or animal nature or origin of the material contained in it, 'natural' or similar words, may be used only if the flavouring components have been isolated solely or almost solely from that vegetable or animal source.

The proposed regulation on flavourings and certain food ingredients with flavouring properties for use in and on foods sets tighter controls for the use of the term 'natural' in the labelling of flavourings which is discussed later in this chapter.

Sweeteners

Foods that contain sweeteners must be labelled with the indication 'with sweetener(s)', and those that contain sugars and sweeteners with the indication 'with sugar(s) and sweetener(s)'. These statements must accompany the product name.

Foods that contain aspartame must be labelled with the words 'contains a source of phenylalanine'.

Foods that contain more than 10% added polyols must carry the indication 'excessive consumption may produce laxative effects'.

Exemptions from Ingredient Listing

Ingredients which need not be named:

i)   constituents of an ingredient which have become temporarily separated during the manufacturing process and are later re-introduced in their original proportions
ii) any additive whose presence in the food is due only to the fact that it was contained in an ingredient of the food, provided it does not serve any significant technological function in the finished product
(iii) any additive that is used solely as a processing aid
(iv) any substance other than water that is used as a solvent or carrier for an additive and is used in an amount that is no more than that which is strictly necessary for that purpose.
(v) Any substance which is not an additive but which is used in the same way and for the same purpose as a processing aid.

However, the allergen labelling requirements detailed later override the above exemptions.

Appropriate durability indication

All foods must be date marked unless specifically exempt from this requirement (see Appendix B for list of exemptions). Highly perishable foods with the potential to endanger human health must be labelled with a 'use by' date, for other foods a minimum durability date must be given.

The date and any storage conditions that need to be observed may be placed apart from the 'best before' or 'use by', as long as there is a reference to the place where the date appears, e.g. 'best before end - see lid'.

Labelling of minimum durability

The words 'best before' must be used to indicate the minimum durability. It must be followed by the date, shown as the day, month and year. For foods that will keep for 3 months or less, the label may state 'best before' with the day and the month only.

Foods that will keep for more than 3 months but not more than 18 months may be labelled with 'best before end' with the month and the year only. For foods that will last longer than 18 months, the label may state 'best before end' plus month and year only, or year only.

Labelling of 'use by' date

If 'use by' is required, it must be followed by the day and month or the day, month and year in that order.

Quantitative Ingredients Declaration (QUID)

The aim of QUID labelling is to help consumers differentiate between similar products and so be able to make a more informed choice.

It is required that the quantity of an ingredient or category of ingredients used in the manufacture or preparation of a foodstuff is declared where the ingredient:

i)   appears in the name under which the food is sold, or is usually associated with that name by the consumer; or
ii) is emphasised in the labelling, either by words or by the use of pictorial representations; or
iii) is essential to characterise the food and to distinguish it from products with which it could be confused because of its name or appearance; or
iv) in other cases, as determined.

QUID is not required if:

i)   the drained net weight of the food is indicated;
ii) the quantity of the ingredient is already required to be given;
iii) the ingredient is used in small quantities for the purpose of flavouring;
iv) the name of the ingredient appears in the name under which the food is sold, but where the variation in its quantity does not distinguish the food from similar products.

The quantity must be expressed as a percentage and must correspond to the quantity of the ingredient at the time of use. The declaration must appear either in or immediately next to the name under which the food is sold or in an appropriate place in the list of ingredients.

QUID declarations are not triggered by:

- 'with sweeteners' or 'with sugars and sweeteners', in the name of the food;
- references to vitamins and minerals, as long as these are indicated in nutrition labelling.

QUID calculations

QUID declarations should be calculated on the finished product for foodstuffs that have lost moisture following heat treatment or other treatment. If the resultant % exceeds 100%, then it is to be replaced by the weight of ingredient used in the preparation of 100 g of the finished product.

In the case of volatile ingredients, QUID should be calculated on the finished product.

Dehydrated or concentrated ingredients, which are reconstituted during manufacture, may be declared on the basis of ingredient weight prior to concentration or dehydration. Alternatively, for concentrated or dehydrated foods that are intended to be reconstituted with water, QUID may be given on the basis of the reconstituted product.

Storage conditions

Any special storage conditions or conditions of use need to be included on the label.

Name and address

The business name and address of the manufacturer or packer, and/or seller established within the European Community needs to be included on the label.

Place of origin

Details of the place of origin of the food must be given if the failure to provide such information would mislead as to the true origin of the food.

Instructions for use

These need to be included if it would be difficult to use the food without instructions.

Additional labelling

There are additional labelling requirements for the following types of food:

Food sold from vending machines
Prepacked alcoholic drinks
Raw milk
Products that contain skimmed milk with non-milk fat
Foods packaged in certain gases

Certain foods with compositional standards (e.g. jams, chocolate, infant formulae) also have additional labelling requirements specified within the appropriate compositional regulation.

There are labelling requirements for some cheese varieties, cream types, milk, ice cream and indication of specific flavours, which aim to prevent misleading descriptions set in Schedule 8 to the Food Labelling Regulations.

Allergens

EC Directive 2000/13/EC as amended states that foods containing allergenic ingredients or ingredients originating from an allergenic ingredient listed below must be marked with a declaration of these ingredients in the ingredients list, unless they have already been mentioned in the product name.

This means that even if an ingredient meets the criteria for which it wouldn't usually need to be declared in an ingredients list, or could have been declared by a generic name, if it contains an allergenic ingredient or originated from one, the ingredient must be declared.

The current list of allergens is:

- Cereals containing gluten: wheat, rye, barley, oats, spelt, kamut and their hybridised strains
- Crustaceans
- Eggs
- Fish
- Peanuts

- Soybeans
- Milk
- Nuts - specific varieties
- Celery
- Mustard
- Sesame seeds
- Sulphur dioxide and sulphites at concentrations of more than 10 mg/kg or 10 mg/l, expressed as SO_2
- Lupin*
- Molluscs*

*Lupin and molluscs were added to the list of allergens under Commission Directive 2006/142/EC and member states were required to transpose this Directive into national legislation by 23 December 2007. For example, in the UK, this Directive is implemented through the Food Labelling (Declaration of Allergens) (England) Regulations 2007.

All products must comply with the requirements of this Directive by 23 December 2008.

Directive 2007/68/EC (amending Annex IIIa to Directive 2000/13/EC) has been published following a review by EFSA of dossiers submitted for highly processed ingredients derived from the allergens listed in Annex IIIa, to allow exemption from labelling with reference to the allergen.

The exemptions are set out below:

(a) wheat-based glucose syrups including dextrose [1];
(b) wheat-based maltodextrins [1];
(c) glucose syrups based on barley;
(d) cereals used for making distillates or ethyl alcohol of agricultural origin for spirit drinks and other alcoholic beverages;
(e) fish gelatine used as carrier for vitamin or carotenoid preparations;
(f) fish gelatine or Isinglass used as fining agent in beer and wine;
(g) fully refined soybean oil and fat [1];
(h) natural mixed tocopherols (E306), natural D-alpha tocopherol, natural D-alpha tocopherol acetate, natural D-alpha tocopherol succinate from soybean sources;
(i) vegetable oils derived phytosterols and phytosterol esters from soybean sources;
(j) plant stanol ester produced from vegetable oil sterols from soybean sources;
(k) whey used for making distillates or ethyl alcohol of agricultural origin for spirit drinks and other alcoholic beverages;

(l) lactitol;

(m) nuts used for making distillates or ethyl alcohol of agricultural origin for spirit drinks and other alcoholic beverages.

[1] And products thereof, insofar as the process that they have undergone is not likely to increase the level of allergenicity assessed by the EFSA for the relevant product from which they originated.

Alcoholic drinks which have an alcoholic strength by volume of more than 1.2% and contain any allergenic ingredient listed need to be labelled with the word 'contains' followed by the name of the allergenic ingredient.

Prescribed nutrition labelling

When a claim is made and/or food is fortified, prescribed nutrition labelling is triggered; otherwise nutritional labelling is voluntary. The only exceptions are natural mineral waters and food supplements, which are exempt from prescribed nutrition labelling as set in the Food Labelling Regulations but are subject to product specific controls. There are several different criteria for nutrition labelling, depending on the type of claim being made.

Prescribed nutrition labelling must include either Group 1 (a) or Group 2 (b):

(a) energy and the amounts of protein, carbohydrate and fat; or

(b) energy and the amounts of protein, carbohydrate, sugars, fat, saturates, fibre and sodium (this format should be used if a claim is being made for sugars, saturates, fibre or sodium).

Where a nutrition claim is made for polyols, starch, monounsaturates, polyunsaturates, cholesterol, vitamins or minerals, the amount/s must be included in the prescribed nutrition labelling. Where no claim is made, these nutrients may be optionally included.

The nutrients need to be listed in the following order and in the same style:

energy	[x] kJ and [x] kcal
protein	[x] g
carbohydrate	[x] g
of which:	
- sugars	[x] g

- polyols	[x] g
- starch	[x] g
fat	
of which:	
- saturates	[x] g
- monounsaturates	[x] g
- polyunsaturates	[x] g
- cholesterol	[x] mg
fibre	[x] g
sodium	[x] g
[vitamins]	[x units]
[minerals]	[x units]

Where monounsaturates and/or polyunsaturates are included, saturates must also be included.

All amounts must be expressed per 100 g or 100 ml of the food. In addition, they may be given per quantified serving of food or per portion of food.

The use of nutrition claims is controlled through Regulation (EC) No. 1924/2006 on nutrition and health claims made on foods which was applied from 1 July 2007 and is directly applicable in England through the Nutrition and Health Claims Regulations 2007. Making a claim is voluntary and the regulation details this information to include conditions for their use. Any foods which do not meet the requirements stated in the regulations would be subject to transitional measures. A health claim triggers group 2 nutrition declaration and other additional labelling requirements.

As well as this, Regulation (EC) 1925/2006 on the Addition of Vitamins and Minerals and of certain other substances to foods requires that foods to which vitamins and minerals have been added (covered by Regulation (EC) 1925/2006) must contain nutrition labelling and be of the Group 2 format as described previously in this section.

At time of publication, EU food labelling legislation is being reviewed by the European Commission; it is likely that the current food labelling directive will be superseded by a directly applicable regulation.

The Commission published a draft proposal for a Regulation on the provision of food information to consumers at the end of 2007. The draft proposal includes a new requirement to provide mandatory nutrition labelling for the energy value and the amount of fats, saturated fats, sugars and salt, and requires this information to be given in the principal field of vision of a food label and in this order. It is anticipated that this regulation will be adopted by the European Parliament and Council by 2010.

The national provisions of the Food Labelling Regulations 1996 (as amended) are being reviewed by the Food Standards Agency with the intention of either removing or seeking to retain specific provisions for inclusion at European level.

Food Additives Labelling Regulations 1992 (S.I. 1992 No. 1978)

These Regulations relate to business and consumer sales of food additives sold as such. They define food additives, list excluded substances including processing aids, a definition of which is given, and prescribe requirements for labelling.
These Regulations do not apply to

i)   Processing aids
ii)   Substances used in the protection of plants and plant products.
iii) Flavourings within the meaning of the Flavourings in Food Regulations 1992.
iv) Substances added to foods as nutrients.

Definition

A food additive must fall within a category or categories listed below:

Colours
Antioxidants
Preservatives
Emulsifiers
Emulsifying salts
Thickeners
Gelling agents
Stabilisers
Flavour enhancers
Acids
Acidity regulators
Anti-caking agents
Modified starch
Sweeteners
Raising agents
Anti-foaming agents
Glazing agents
Flour bleaching agents: any substance primarily used to remove colour from flour

Flour treatment agents: any substance that is added to flour or dough to improve its baking quality
Firming agents
Humectants
Enzyme preparations: any substance that contains a protein capable of catalysing a specific chemical reaction
Sequestrants
Bulking agents
Propellants
Packaging gas
Carriers and carrier solvents

An additive is normally neither consumed as a food in itself or used as a characteristic ingredient of food, whether or not it has nutritive value, and is intentionally added to food for a technological purpose in the manufacture, processing, preparation, treatment, packaging, transport or storage of that food, and results, or could result, in it or its by-products becoming directly or indirectly a component of the food.

A processing aid is defined as a substance that is not consumed as a food ingredient by itself; is intentionally used in the processing of raw materials, foods or their ingredients, to fulfil technological purposes during treatment or processing; and is capable of resulting in the unintended but technically unavoidable presence of its residues or its derivatives in the finished product, and the residues of which do not present any risk to human health and do not have any technological effect on finished products.

Labelling requirements for business sale of food additives

The container of the food additive must bear the information listed under 1 or 2 below, and it must be clearly legible, conspicuous and indelible.

1.
(a) The label must have the correct EC name and number, or in the absence of such name, a description of the food additive that will distinguish it from any other that it could be confused with. If there is more than one food additive present, the information must be given in descending order of the proportion by weight.
(b) If there is any supplementary material, (substances to facilitate storage, sale, standardisation, dilution or dissolution of a food additive), each component

of the supplementary material must be labelled in descending order of the proportion by weight of the components.

(c) The label must state that the food additives are 'for use in food' or 'restricted use in food' or a more specific reference to its intended food use.

(d) If there are any special storage conditions for the food additive, or if there are any special conditions of use, this needs to be labelled.

(e) Instructions for the use of the food additive must be given if it would be difficult to use the food additive without them.

(f) An identifying batch or lot mark.

(g) The name and address of the manufacturer or packer, or EC seller of the food additive must be stated on the label.

(h) If it is prohibited to exceed a specified quantity of the food additive in a food, the percentage of each component of the food additive must be stated. Alternatively, enough information must be given to enable the purchaser to decide whether, and to what level, he could use such food additives in food sold by him.

or

2. The label needs to include 1(a), (c), (d) and (e) (above) and in an obvious place the words 'intended for manufacture of foodstuffs and not for retail sale'. Relevant trade documents must be supplied to the purchaser and must include the remainder of the information given in section 1(b), (f), (g) and (h).

Labelling requirements for consumer sale of food additives

The container must bear the following information, which must be clearly legible, conspicuous and indelible.

(a) The name of the product. A description of food additives specified in Community provisions and the EC number. If there is no EC name or EC number, a description must be given to identify it from any other product with which it may be confused.

(b) In addition, the label must include all the information stated in section 1. (a)–(g) in 'Labelling requirements for business sale', above.

(c) The minimum durability of the product must be stated.

Exemption:

These regulations do not apply to any food additive that is part of another food.

Additive numbers

Where the serial number of the additive is to be given in the ingredients list:

- The number used should be one that appears in the column headed 'EC No.' in the relevant schedule (e.g. E150b, E420).

Additive names

Where the specific name of the additive is to be given in the ingredients list:

- The name used should be one that appears in the column headed 'Colour' or 'Permitted sweetener' or 'Name' in the relevant schedule (e.g. Cochineal, Aspartame).
- A summary name that appears in the column headed 'Colour' or 'Permitted sweetener' or 'Name' in the relevant schedule may be used in place of a more specific name, provided that the latter does not have its own serial number (e.g. carotene may be used for 'mixed carotenes', 'sorbitol' may be used for 'sorbitol syrup').
- If the name in the column headed 'Colour' or 'Permitted sweetener' or 'Name' in the relevant schedule is preceded by a bracketed letter or Roman numeral (e.g. (ii) Beta carotene), this need not be given as part of the name.
- In the case of miscellaneous additives, where an alternative to the specific name is given in brackets in the column headed 'Name' in the relevant schedule, this may be used in place of the specific name (e.g. 'polysorbate 20' instead of 'polyoxyethylene sorbitan monolaurate').
- In the case of miscellaneous additives being phosphates, the names, 'diphosphates', 'triphosphates' and 'polyphosphates' are acceptable as specific names for the phosphates covered by the serial numbers E450, E451 and E452, respectively. They should not be used for the phosphates covered by serial numbers E338, E339, E340 and E341.
- Synonyms or acronyms that are not included in the relevant schedule should not be used as alternatives to the specific name.

Relevant schedules

Schedule 1 to the Sweeteners in Food Regulations 1995 as amended by the Sweeteners in Food (Amendment) Regulations 1997
Schedule 1 to the Colours in Food Regulations 1995
Schedules 1, 2, 3, and 4 of the Miscellaneous Food Additives Regulations 1995, as amended.

Flavourings in Food Regulations 1992 (S.I. 1992 No. 1971, as amended by S.I. 1994 No. 1486)

Definition

A flavouring is a material used or intended for use in or on food to impart odour, taste or both.

Labelling requirements for business sale of relevant flavourings

The container must be labelled with the following information.

(a) The name and business name and address of the manufacturer or the packer, or of the EC seller.
(b) The word 'flavouring' or more specific names or descriptions of the relevant flavourings.
(c) The words 'for foodstuffs' or a more specific reference to the food for which the relevant flavouring is intended.
(d) A list, in descending order of weight, using the following classifications:

  - 'natural flavouring substances' – for flavouring substances obtained by physical, enzymatic or microbiological processes from appropriate material of vegetable or animal origin;
  - 'flavouring substances identical to natural substances' – for flavouring substances obtained from chemical synthesis or isolated by chemical processes and chemically identical to a substance naturally present in appropriate material of vegetable or animal origin;
  - 'artifical flavouring substances' – for flavouring substances obtained by chemical synthesis;
  - 'flavouring preparations' – for flavouring preparations;
  - 'process flavourings' - for process flavourings;
  - 'smoke flavourings' - for smoke flavourings.

In the case of other substances or materials, their names or EC numbers.

(e) The quantity of any material in or on the relevant flavourings where the sale of food containing excess of such quantity would be prohibited by the Food Safety Act.

The information must be visible, legible, and indelible, and must be expressed in terms easily understood by the purchaser.

Use of the word 'natural'

The word 'natural' or any similar word may not be used to describe the relevant flavouring unless it is used in compliance with the labelling requirements above; or the flavouring components of the relevant flavouring comprise flavouring substances obtained by physical, enzymatic or microbiological processes from appropriate material of vegetable or animal origin or flavouring preparations or both.

The word 'natural' or any similar word shall not be used to qualify any substance used in its preparation unless the relevant flavouring is a permitted flavouring, the flavouring component of which has been isolated solely, or almost solely, from that substance by physical processes, enzymatic or microbiological processes, or processes normally used in preparing food for human consumption.

The conditions governing the use of the word 'natural flavouring' in labelling will be amended by the proposed regulation on flavourings and certain food ingredients with flavouring properties for use in and on foods as discussed later in this chapter. According to this proposed regulation, the term 'natural' may only be used for the description of flavouring if the flavouring component comprises only flavouring preparations and/or natural flavouring substances.

A 'natural flavouring substance' shall mean a flavouring substance obtained by appropriate physical, enzymatic or microbiological processes from material of vegetable, animal or microbiological origin either in the raw state or after processing for human consumption by one or more of the traditional food preparation processes as listed in the regulation. They are:

Chopping
Coating
Cooking, baking, frying (up to 240 °C)
Cooling
Cutting
Distillation / rectification

Drying
Emulsification
Evaporation
Extraction, including solvent extraction
Fermentation
Filtration
Grinding
Heating
Infusion
Maceration
Microbiological processes
Mixing
Peeling
Percolation
Pressing
Refrigeration/freezing
Roasting/grilling
Squeezing
Steeping

For 'flavouring preparation' it is natural under the conditions that it is a product, other than a flavouring substance which is obtained from food by appropriate physical, enzymatic or microbiological processes either in the raw state of the material or after processing for human consumption by one or more of the traditional food preparation processes listed above and/or appropriate physical processes.

The term 'natural' may only be used in combination with a reference to a food, food category or a vegetable or animal flavouring source, if at least 95% (by w/w) of the flavouring component has been obtained from the source material referred to.

The flavouring component may contain flavouring preparations and/or natural flavouring substances.

Labelling requirements for consumer sale of relevant flavourings

The container must include the following information:

(a) The name and business name and address of the manufacturer or the packer, or of the EC seller.

(b) The word 'flavouring' or more specific names or descriptions of the relevant flavourings.
(c) The words 'for foodstuffs' or a more specific reference to the food for which the relevant flavouring is intended.
(d) An indication of minimum durability.
(e) Any special storage conditions or conditions of use.
(f) Instructions for use, where omission would prevent appropriate use of the flavouring.
(g) Where the relevant flavouring contains other substances or materials, a list in descending order of weight:

- in respect of components of the relevant flavouring, the word 'flavouring' or more specific names or descriptions of the relevant flavourings;
- in respect of each other substance or material, its name or, where appropriate, its E number.

The information must be visible, legible and indelible, and must be expressed in terms easily understood by the purchaser.

Sale of food containing flavourings:
Generally no food shall be sold which has in it or on it any added relevant flavouring other than a permitted flavouring (complying with general purity criteria).

Relevant schedule:
Schedule 1: General purity criteria applicable to permitted flavourings.

Likewise no food sold which has in it or on it any relevant flavouring shall have in it or on it any specified substance which has been added as such.
These specified substances may be present in a food either naturally or as a result of the inclusion of the relevant flavouring which has been made from natural raw materials.
The presence of the specified substance in foods that are ready for consumption should not exceed specified limits.

Relevant schedule:
Schedule 2: Specified substances

Smoke Flavourings Regulations 2005 (S.I. 2005 No. 464)

Definition

'Smoke flavouring' means a smoke extract used in traditional foodstuffs smoking processes.

The following definitions are also given:

- 'primary smoke condensate' shall refer to the purified water-based part of condensed smoke and shall fall within the definition of 'smoke flavourings';
- 'primary tar fraction' shall refer to the purified fraction of the water-insoluble high-density tar phase of condensed smoke and shall fall within the definition of 'smoke flavourings';
- 'primary products' shall refer to primary smoke condensates and primary tar fractions;
- 'derived smoke flavourings' shall refer to flavourings produced as a result of the further processing of primary products and which are used or intended to be used in or on foods in order to impart smoke flavour to those foods.

Smoke flavourings need to be indicated as such, see the previous section on Flavourings in food.

Colours in Food Regulations 1995 (S.I. 1995 No. 3124, as amended by S.I. 2000 No. 481, S.I. 2001 No. 3442, S.I. 2005 No. 519 and S.I. 2007 No. 453)

Definition

A food colour is a food additive used or intended to be used primarily for adding or restoring colour to a food. This includes:

- any natural constituent of food and any natural source not normally consumed as food as such and not normally used as a food ingredient; and
- any preparation obtained from food or any other natural source material by physical and/or chemical extraction resulting in selective extraction of the pigment relative to the nutritive or aromatic constituent.

For labelling of colours in foods see the Food Labelling Regulations - Additives section. See section on Food Additives Labelling Regulations, for business/consumer sale of food additives.

Sweeteners in Food Regulations 1995 (S.I. 1995 No. 3123, as amended by S.I. 1996 No. 1477, S.I. 1997 No. 814, S.I. 1999 No. 982, S.I. 2001 No. 2294, S.I. 2002 No. 379, S.I. 2003 No. 1182, S.I. 2004 No. 3348 and S.I. 2007 No. 1778)

Definition

A sweetener is a food additive used or intended to be used to impart a sweet taste to food, or as a table-top sweetener.

For labelling of sweeteners, see sections on:

i)   Food Additives Labelling Regulations, for business/consumer sale of food additives.
ii) The Food Labelling Regulations - both Additives and Sweeteners.

Miscellaneous Food Additives Regulations 1995 (S.I. 1995 No. 3187, as amended by S.I. 1997 No. 1413, S.I. 1999 No. 1136, S.I. 2001 No. 60, S.I. 2001 No. 3775, S.I. 2003 No. 1008, S.I. 2003 No. 3295, S.I. 2004 No. 2601, S.I. 2005 No. 1099 and S.I. 2007 No. 1778)

Definition

The term 'miscellaneous additive' refers to any food additive that is used or intended to be used primarily as an acid, acidity regulator, anti-caking agent, anti-foaming agent, antioxidant, bulking agent, carrier, carrier solvent, emulsifier, emulsifying salt, firming agent, flavour enhancer, flour treatment agent, foaming agent, gelling agent, glazing agent, humectant, modified starch, packaging gas, preservative, propellant, raising agent, sequestrant, stabiliser or thickener; but does not include use as a processing aid or any enzyme except invertase or lysozyme.

Relevant schedules

The Regulations contain nine schedules:

Schedule 1: Miscellaneous additives generally permitted for use in foods
Schedule 2: Conditionally permitted preservatives and antioxidants
Schedule 3: Other permitted miscellaneous additives
Schedule 4: Permitted carriers and carrier solvents
Schedule 5: Purity criteria

Schedule 6: Foods in which miscellaneous additives listed in Schedule 1 are generally prohibited

Schedule 7: Foods in which a limited number of miscellaneous additives listed in Schedule 1 may be used

Schedule 8: Miscellaneous additives permitted in foods for infants and young children

Schedule 9: Revocations

The Regulations function in the form of positives lists, as detailed in the schedules above.

Extraction Solvents in Food Regulations 1993 (S.I. 1993 No. 1658, as amended by S.I. 1995 No. 1440 and S.I. 1998 No. 2257)

Definition

An extraction solvent is any solvent used or intended to be used in an extraction procedure, including, in any particular case further to its use in such a procedure, any substance other than such a solvent derived exclusively from such a solvent.

See Appendix C for the list of Permitted Extraction Solvents.

Labelling of permitted extraction solvents sold as such

The following information must be provided with any of the extraction solvents listed in Appendix C.

a) The name of the permitted extraction solvent that is stated in the list of Permitted Extraction Solvents.

b) A clear statement that the permitted extraction solvent is of suitable quality for use in an extraction procedure.

c) An identifying batch or lot mark.

d) The name or business name and address of the manufacturer or packer, or of an established EC seller.

e) The net quantity or volume, in metric units, of the permitted extraction solvent in any container or other packaging in which it is to be sold or imported.

f) Any special storage conditions or conditions of use.

The information must be easily visible, clearly legible and indelible.

The information must be given on the packaging, container or label of the permitted extraction solvent to which it relates; alternatively, statements c – f may be specified on relevant trade documents that accompany or precede the delivery.

The quantity may also be accompanied with other units of measurement, provided the metric indication is predominant and expressed in characters that are no smaller than the other units.

The Genetically Modified Food (England) Regulations 2004 (S.I. 2004 No. 2335)

This Regulation makes provisions for the enforcement of EC Regulation No. 1829/2003 on genetically modified food and feed which harmonises procedures for the scientific assessment and authorisation of genetically modified organisms (GMOs) and genetically modified food and feed and lays down labelling requirements.

The EC Regulation applies to the whole of the UK although the S.I. 2004 No. 2335 applies only in England. Similar legislation has been made in Scotland, Wales and Northern Ireland as follows:

- The Genetically Modified Food Regulations (Northern Ireland) 2004 (Statutory Rule 2004 No. 385)
- The Genetically Modified Food (Scotland) Regulations 2004 (Scottish Statutory Instrument 2004 No. 432)
- The Genetically Modified Food (Wales) Regulations 2004 No. 3220 (W.276)

The Genetically Modified and Novel Foods (Labelling) (England) Regulations 2000 have been revoked.

The Food Standards Agency was designated as the national competent authority to receive applications for the authorisation of:

- new genetically modified organisms for food use
- food containing or consisting of GMOs or
- food produced from or containing ingredients produced from GMOs.

The Genetically Modified food (England) Regulations sets labelling requirements for:

- food containing or consisting of genetically modified organisms (GMOs) or
- food produced from or containing ingredients produced from GMOs.

The labelling requirements apply regardless of whether or not the final product contains DNA or protein resulting from genetic modification.

The labelling requirements of this regulation do not apply to foods containing material which contains, consists of or is produced from GMOs that have an EU authorisation in a proportion <0.9% of the food ingredients, where the presence is adventitious or technically unavoidable. This unintentional presence is subject to the operator being able to supply evidence to satisfy the competent authorities that they have taken appropriate steps to avoid the presence of such material.

GM material that has not been authorised in the EU cannot be present at any level in food products.

The Regulation requires that:

- If the food consists of more than one ingredient, the words 'genetically modified' or 'produced from genetically modified X', are to appear in the ingredient list in parentheses immediately after the ingredient name (or in a prominent footnote linked to indicate this) in the ingredients list.
- If the ingredient is designated by a category name, the words 'contains genetically modified Y', or 'contains X produced from genetically modified Y', are to appear in the ingredients list.
- For a food without an ingredient list, the words 'genetically modified' or 'produced from genetically modified Y' are to appear on the label.

where, X = name of ingredient, where Y = name of organism

In the case of non-prepackaged products or pre-packaged food in small containers of which the largest surface has an area of less than 10 cm^2, the information required under this paragraph must be permanently and visibly displayed either on the food display or immediately next to it, or on the packaging material, in a font sufficiently large for it to be easily identified and read.

In addition to the labelling requirements given above, the labelling should also mention any characteristic or property, as specified in the authorisation where a food or ingredient has changed in respect to its composition, nutritional value/nutritional effects, intended use, implications for the health of certain sections of the population as well as any ethical/religious concerns.

Also, genetically modified food must not:

- have adverse effects on human health, animal health or the environment
- mislead the consumer

-   differ from the food which it is intended to replace to such an extent that its normal consumption would be nutritionally disadvantageous for the consumer.

Note: Legally there are no controls for 'GM free' labelling for food ingredients other than rules on misleading claims. Consumers should check with the company/retailer as to the criteria that are being employed in using the term.

The Genetically Modified Organisms (Traceability and Labelling) (England) Regulations 2004 (S. I. 2004 No. 2412)

This Regulation provides for the enforcement in England of EC Regulation No. 1830/2003 on the traceability and labelling of GMOs and GM food and feed. It requires the identification of GM products throughout the supply chain in order to facilitate accurate labelling in accordance with Regulation (EC) 1829/2003 for:

-   food consisting or containing GMOs
-   food produced from GMOs

Unique identifier codes on GMOs can be found in a register and are used in traceability documentation. These codes must be used for the traceability of products consisting of or containing GMOs (e.g. maize) but not of foods produced from GMOs (e.g. maize gluten).

The EC Regulation applies to the whole of the UK and similar legislation has been made in Scotland, Wales and Northern Ireland.

Package of proposals for new legislation on food additives, flavourings and enzymes

In July 2006, The European Commission published a package of legislative proposals to introduce harmonised EU legislation on food enzymes for the first time and upgrade current rules for food flavourings and additives to bring them into line with the latest scientific and technological developments. The proposals were amended in October 2007.

The package includes four proposals on food improvement agents as follows:

1. Establishing a common authorisation procedure for food additives, food enzymes and food flavourings
2. Food additives

3. Food enzymes
4. Flavourings and certain food ingredients with flavouring properties for use in and on foods

1. Proposal for a Regulation of the European Parliament and of the Council establishing a common authorisation procedure for food additives, food enzymes and food flavourings
http://eur-lex.europa.eu/LexUriServ/site/en/com/2007/com2007_0672en01.pdf

The safety of additives, enzymes and flavourings used in foodstuffs for human consumption must be assessed before they can be placed on the community market.

Currently, the general criteria for the use of food additives is given in the Framework Directive 89/107/EEC concerning food additives authorised for use in foodstuffs intended for human consumption, which is discussed further in the next chapter. The authorisation procedure for a food additive at Community level currently involves a two-step procedure. Therefore, firstly the additive is included in the relevant Directive, and then the Commission would adopt a specification for that additive after this is agreed by the Standing Committee on the Food Chain and Animal Health.

The proposed Regulation lays down a common assessment and authorisation procedure for food additives, food enzymes, food flavourings and sources of food flavourings used or intended for use in or on foodstuffs.

Under Regulation (EC) 178/2002 laying down procedures in matters of food safety, the placing of substances on the market must be authorised only after an independent scientific assessment by the European Food Safety Authority of the risks that they pose to human health. This is followed by a risk management decision taken by the Commission. Food additives, food enzymes, food and flavourings must be included in the positive list for each respective regulation in order for them to be marketed for human consumption. These positive lists will be created, maintained and published by the Commission.

2. Proposal for a Regulation of the European Parliament and of the Council on food additives
http://eur-lex.europa.eu/LexUriServ/site/en/com/2007/com2007_0673en01.pdf

Currently, food additives are governed by the following:

• Council Directive 89/107/EEC of 21 December 1988 on the approximation of the laws of the Member States concerning food additives authorised for

use in foodstuffs intended for human consumption. Official Journal of the European Communities L40, 11.02.89,27- 33, as amended.
- European Parliament and Council Directive 94/35/EC of 30 June 1994 on sweeteners for use in foodstuffs. Official Journal of the European Communities. L237, 10.9.94, 3-12, as last amended.
- European Parliament and Council Directive 94/36/EC of 30 June 1994 on colours for use in foodstuffs. Official Journal of the European Communities. L237, 10.9.94, 13-29.
- European Parliament and Council Directive 95/2/EC of 20 February 1995 on food additives other than colours and sweeteners. Official Journal of the European Communities. L61, 18.3.95, 1-40, as last amended.
- Decision No 292/97/EC of the European Parliament and of the Council of 19 December 1997 on the maintenance of national laws prohibiting the use of certain additives in the production of certain specific foodstuffs

At present, the authorisation of a food additive at community level is based on a co-decision procedure. If the new proposal is adopted, the provisions on additives in the different existing Directives will be brought together in one regulation. This single Regulation will harmonise the use of food additives in foods in the Community.

The regulation will include the principles for the use of food additives, and a positive list of approved food additives, as well as covering the use of food additives in food additives and food enzymes, and carriers for nutrients. The regulation will be based on a comitology approach since the package was adopted around the time of entry into force of Decision 2006/512/EC, amending Decision 1999/468/EC laying down the procedures for the exercise of implementing powers conferred on the Commission.

The inclusion of food additives onto a positive list will be based on their safety when used, a technological need and their usage must be of benefit to the consumer. Their use must not mislead the consumer and this would include issues related to the quality of ingredients used, naturalness, nutritional quality of the product or its fruit and vegetable content. The European Food Safety Authority (EFSA) will be responsible for carrying out all safety evaluations.

The additives present in the positive list will have specifications including purity criteria and origin.

Producers or users of additives should provide the Commission with information on their use which may affect the assessment of the safety of the food additive.

When a food additive is already included in a Community list but there is a significant change in the production methods or the starting materials, the food additive prepared by these new methods or materials shall be considered as a different additive and a new entry in the Community lists or change in the specifications shall be required before it can be placed on the market.

Additionally, any GM-containing additives must be authorised following Regulation (EC) No 1829/2003 on genetically modified food and feed.

Food Additives currently included in Directives 95/2/EC, 94/35/EC and 94/36/EC will be entered into Annex II of the proposal following a review undertaken by the Standing Committee on Food Chain and Animal Health (SCFCAH). The SCFCAH will evaluate the compliance of existing authorisations for food additives and their conditions of use with general criteria i.e. technological needs, and consumer aspects. However, Annex III will be completed with other food additives used in food additives and food enzymes as well as carriers for nutrients and their conditions for use as follows:

Annex III:

Part I: Carriers in food additives (transferred from Annex V of Directive 95/2/EC on food additives authorised for use in food additives as permitted carriers/carrier solvents)

Part 2: Additives other than carriers in food additives

Part 3: Additives including carriers in food enzymes

Part 4: Additives including carriers in food flavourings (transferred from Directive 95/2/EC on food additives authorised for use in food flavourings)

Part 5: Carriers in nutrients

3. Proposal for a Regulation of the European Parliament and of the Council on food enzymes
http://eur-lex.europa.eu/LexUriServ/site/en/com/2007/com2007_0670en01.pdf

Currently, Council Directive 95/2/EC on food additives other than colours and sweeteners allows two enzymes to be used as food additives. (In addition, Council Directive 2001/112/EC relating to fruit juices and certain similar products intended for human consumption, Council Directive 83/417/EEC relating to certain lactoproteins intended for human consumption and Council Regulation 1493/1999/EC on the common organisation of the market in wine, regulate the use of certain food enzymes in these specific foods.) Other uses of enzymes aren't regulated at all or are regulated as processing aids under the national legislation of the Member States; however, requirements differ significantly between each Member State.

The proposed regulation on enzymes would establish a positive list of approved enzymes, conditions for their use in foods and rules on labelling of food enzymes sold as such.

Hence, the regulation will apply to all enzymes including enzymes used as processing aids and miscellaneous additives, although the regulation will not apply to enzymes for nutritional or digestive purposes. Likewise, microbial cultures traditionally used in the production of food (e.g. cheese) that may contain enzymes but aren't specifically used to make them will not be considered as food enzymes.

The proposed regulation also defines the terms 'enzyme', 'food enzyme' and 'food enzyme preparation'.

In the positive list, the entry of a food enzyme shall specify:

- The description of the food enzyme (including its common name)
- Specification (including origin, purity criteria etc)
- Foods in which it may be used
- Conditions for its use
- If there any restrictions for the enzyme when sold directly to consumers
- Any specific labelling requirements (in the food where the enzyme has been used to ensure the physical condition of the food and specific treatment is indicated if necessary)

The proposed regulation lays down labelling requirements of food enzymes and food enzyme preparations whether or not they are intended for sale to the final consumer.

Enzymes that are already on the market can be transferred onto the positive list if EFSA accepts the previous safety assessment done at community level. The proposal states there is an initial two-year authorisation period during which EFSA must evaluate all applications for food enzymes.

Novel foods falling within the scope of Regulation (EC) No 258/97 concerning novel foods and novel food ingredients should be excluded from the scope of this proposed regulation on food enzymes.

Enzymes produced from genetically modified organisms will be subject to the scope of Regulation (EC) 1829/2003 on genetically modified food and feed in relation to the safety assessment of the genetic modification, whereas other aspects of safety and the final authorisation shall be covered under the proposed regulation on food enzymes.

4. Proposal for a Regulation of the European Parliament and of the Council on flavourings and certain food ingredients with flavouring properties for use in and on foods
http://eur-lex.europa.eu/LexUriServ/site/en/com/2007/com2007_0671en01.pdf

Currently, flavourings are regulated through Council Directive 88/388/EEC relating to flavourings for use in foodstuffs and to source materials for their production. Within this, flavourings can be divided into the following categories:

- Flavouring substances (which describes natural, nature identical and artificial flavourings)
- Flavouring preparations
- Process flavourings
- Smoke flavourings

Council Directive 88/388/EEC also sets maximum limits for certain undesirable substances obtained from flavourings and other food ingredients with flavouring properties. The proposed regulation on flavourings sets new maximum limits for the presence of these undesirable substances in foods. It also introduces a new annex (Annex IV) which lists source materials to which restrictions apply for their use in the production of flavourings and food ingredients with flavouring properties. EFSA is responsible for the risk assessment of flavourings.

The proposed regulation on flavourings aims to establish a positive list of flavourings and source materials approved for use in and on foods with their conditions of use in and on foods, as well as setting rules on the labelling of flavourings.

The positive list shall be established by placing the list of flavouring substances referred to in Article 2(2) of Regulation (EC) No 2232/96 laying down a Community procedure for flavouring substances used or intended for use in or on foodstuffs, into Annex I of the proposed regulation on flavourings, at the time of its adoption.

The current register of flavouring substances (as adopted by Decision (EC) 1999/217/EC, as amended, adopting a register of flavouring substances used in or on foodstuffs drawn up in application of Regulation (EC) No 2232/96), is valid for the whole of the EU and includes about 2700 flavouring substances.

The new proposed regulation defines and contains the following categories of flavourings:

- Flavouring substances (defining natural flavouring substances only)
- Flavouring preparations

- Thermal process flavourings*
- Smoke flavourings
- Flavour precursors*
- Other flavourings or mixtures thereof*

*A new category of flavouring introduced by the proposal

The definition of a 'natural flavouring substance' has been amended by this proposal as discussed earlier in this chapter. Given that the chemical structure of the molecules is identical, it was sensible to remove the distinction between 'natural' and 'nature identical' flavouring substances because as far as human consumption is concerned, it is the safety of the substance that is important, not its origin.

A 'thermal process flavouring' is defined as a product obtained after heat treatment from a mixture of ingredients not necessarily having flavouring properties themselves, of which at least one contains nitrogen and another is a reducing sugar; the ingredients for the production of thermal process flavourings may be:

(i) food; and/or
(ii) source material other than food

In Annex V of the proposed regulation on flavourings, conditions for the production of thermal process flavourings and maximum levels for certain substances in thermal process flavourings are specified.

A 'flavour precursor' is defined as a product, not necessarily having flavouring properties itself, intentionally added to food for the sole purpose of producing flavour by breaking down or reacting with other components during food processing; it may be obtained from:

(i) food; and/or
(ii) source material other than food;

A flavouring or source material that falls within the scope of Regulation (EC) 1829/2003 on genetically modified food and feed can only be included in the positive list of flavourings under the new proposal if it has been covered by an authorisation in accordance with the Regulation (EC) No 1829/2003.

Bibliography

Leatherhead Food International. Guide to Food Regulations in the United Kingdom. Leatherhead, Leatherhead Food International Ltd. 1995 (updated to October 2007).

O'Rourke R. European Food Law. Isle of Wight, Palladian Law Publishing. 1998.

Food Standards Agency. Guidance Notes on Nutrition Labelling. 1999.

Food Standards Agency. Guidance Notes on What Foods Should Carry a 'Use by' Date. 2003.

Food Standards Agency. Guidance Notes on Quantitative Ingredient Declarations. 1999.

Directive 2005/29/EC of the European Parliament and of the Council of 11 May 2005 concerning unfair business-to-consumer commercial practices in the internal market and amending Council Directive 84/450/EEC, Directives 97/7/EC, 98/27/EC and 2002/65/EC of the European Parliament and of the Council and Regulation (EC) No 2006/2004 of the European Parliament and of the Council (Unfair Commercial Practices Directive).

The Nutrition and Health Claims (England) Regulations 2007 (S.I. 2007 No. 2080).

Commission Directive 2007/68/EC of 27 November 2007 amending Annex IIIa to Directive 2000/13/EC of the European Parliament and of the Council as regards certain food ingredients.

The Food Labelling (Declaration of Allergens) (England) Regulations 2007 (S.I. 2007 No. 3256).

Regulation (EC) No 1829/2003 of the European Parliament and of the Council of 22 September 2003 on genetically modified food and feed.

Regulation (EC) No 1830/2003 of the European Parliament and of the Council of 22 September 2003 concerning the traceability and labelling of genetically modified organisms and the traceability of food and feed products produced from genetically modified organisms and amending Directive 2001/18/EC.

Appendix A: Exemptions from Food Ingredients Listing

Foods that are exempt from ingredients listing are:

1. Fresh fruit and vegetables, which have not been peeled or cut into pieces.
2. Carbonated water that contains only carbon dioxide, and whose name indicates that it has been carbonated.
3. Vinegar obtained by fermentation from a single product with no additions.
4. Cheese, butter, fermented milk and fermented cream containing only lactic products, enzymes and microorganism cultures essential to manufacture, or cheese (except curd cheese and processed cheese) containing salt for manufacture.
5. Flour to which no substances have been added other than those required to be present in flour by the Bread and Flour Regulations 1998.
6. Drinks with an alcoholic strength by volume of more than 1.2%.
7. Foods consisting of a single ingredient, where the name of the food is identical to the name of the ingredient, or the name of the food enables the nature of the ingredient to be clearly identified.

Appendix B: Exemptions from Durability Indication

1. Fresh fruit and vegetables, which have not been peeled or cut into pieces.
2. Wine, liqueur wine, sparkling wine, aromatised wine and any similar drink obtained from fruit other than grapes.
3. Any drink made from grapes or grape musts and coming within specified codes of the Combined Nomenclature.
4. Any drink with an alcoholic strength by volume of 10% or more.
5. Any soft drink, fruit juice or fruit nectar or alcoholic drink, sold in a container containing more than 5 litres and intended for supply to catering establishments.
6. Any flour confectionery and bread that, given the nature of its content, is normally consumed within 24 hours of its preparation.
7. Vinegar.
8. Cooking and table salt.
9. Solid sugar and products consisting almost solely of flavoured or coloured sugars.
10. Chewing gums and similar products.
11. Edible ices in individual portions.

Appendix C: Permitted Extraction Solvents

1. Propane
2. Butane
3. Ethyl acetate
4. Ethanol
5. Carbon dioxide
6. Acetone
7. Nitrous oxide
8. Methanol
9. Propan-2-ol
10. Hexane
11. Methyl acetate
12. Ethylmethylketone
13. Dichloromethane
14. Diethyl ether
15. Butan-1-ol
16. Butan-2-ol
17. Propan-1-ol
18. Cyclohexane
19. 1,1,1,2-Tetrafluoroethane

3. SAFETY OF FOOD ADDITIVES IN EUROPE

Introduction

The objective of European Union (EU) legislation on food additives is to ensure protection of public health within a harmonised EU internal food market. The legislation on food additives has been developed following the approach laid down by the European Commission in 1985 (1). This approach limited the requirement for legislation to those areas that were justified by the need to protect public health, to provide consumers with information and protection in matters other than health, to ensure fair trading and to provide for the necessary public controls. This chapter focuses on the mechanisms to ensure the safety of food additives covered by EU legislation.

European Directives

Framework Directive on Food Additives

The general framework Directive 89/107/EEC on food additives was adopted by the European Economic Community in 1988 (2). This Council Directive:-

- gives a definition of 'food additive'
- sets out a framework for adoption of lists of permitted additives
- gives general criteria for the inclusion of food additives on such lists
- provides for the adoption of purity criteria (specifications)
- gives Member States powers to temporarily suspend or restrict the use of a permitted additive if new information gives grounds for thinking it might endanger health
- gives Member States powers to provisionally authorise new additives at the national level for up to 2 years
- provides for consultation of the Scientific Committee on Food or, since May 2003 by its replacement the European Food Safety Authority (EFSA) in matters concerning public health,
- provides for labelling of traded food additives and of foods sold to the consumer.

The definition of a food additive in 89/107/EEC is as follows:-

"Any substance not normally consumed as a food in itself and not normally used as a characteristic ingredient of food whether or not it has nutritive value, the intentional addition of which to food for a technological purpose in the manufacture, processing, preparation, treatment, packaging, transport or storage of such food results, or may be reasonably expected to result, in it or its by-products becoming directly or indirectly a component of such foods".

The definition excludes processing aids, including enzymes and extraction solvents, flavourings, substances added as nutrients, such as vitamins and minerals, and substances migrating from food-packaging materials that do not exert a technological function in the food. All substances falling under this definition are called food additives in the EU, and there is no distinction, as there is in the USA, into 'direct' and 'indirect' food additives. Substances defined as food additives in the EU are the equivalent of direct additives in US terminology. Indirect food additives in the US are pesticide residues and substances derived from food-packaging materials. Pesticides and food-packaging substances are covered by separate legislation in the EU and will not be further discussed here. Similarly, extraction solvents and flavourings are also covered by separate EU legislation and will not be further discussed.

Specific Directives on classes of additives

Between 1994 and 1995, three specific Directives stemming from 89/107/EEC were adopted (3-5). They are widely known as the "sweeteners Directive", the "colours Directive" and the "miscellaneous additives Directive". These Directives and their subsequent amendments list the individual permitted additives (now 15 sweeteners, 42 colours and over 280 miscellaneous additives) and the general and specific food categories in which each additive is permitted, and lay down any necessary maximum levels of use. Additives are also grouped into Annexes in the Directives, which broadly define how widely they may be used. All three Directives also require Member States to set up systems to monitor consumer consumption of additives and, in the case of sweeteners, to establish consumer surveys that will include monitoring of 'table-top' sweetener usage (6). Results of such monitoring are to be reported to the Commission and ultimately to the European Parliament.

The new proposal for an EC Regulation on food additives as discussed in the previous chapter would bring together the framework directive, colours, sweeteners and miscellaneous additives directives into one regulation. Therefore once the proposed regulation on food additives is in force, the framework directive, colours, sweeteners and miscellaneous additives directives will be repealed.

Origin of 'E' numbers

Each permitted additive is assigned an 'E' number, signifying that it has been approved as safe for food use by the EC Scientific Committee on Food (SCF), or, since May 2003 by its replacement EFSA, and its inclusion in the relevant directive has been agreed by the Member States. Each E number has a separate specification, which lays down purity criteria for the additive (7-9). Labels on processed foods may list additives by their E numbers and/or by their common name.

Safety Testing and Evaluation of Food Additives

Requirements of the EC Framework Directive on safety assessment

The general criteria for use of food additives set out in Directive 89/107/EEC stipulate that additives can be approved only if they present no hazard to the health of the consumer at the level of use proposed, so far as can be judged on the scientific evidence available (2). To assess the possible harmful effects of a food additive or its derivatives, it must be subject to toxicological testing. All food additives must be kept under observation after approval so that they can be re-evaluated if there are changing conditions of use, or if new scientific information emerges on safety aspects.

Under the new proposal for an EC Regulation on food additives and that of the proposal for an EC regulation establishing a common authorisation procedure for food additives, food enzymes and food flavourings as discussed in the previous chapter, EFSA will be responsible for carrying out all safety evaluations for food additives, whilst the Commission will create, maintain and publish the positive lists for food additives, food enzymes and food flavourings.

General approach of advisory and regulatory bodies

The safety assessment of food additives has developed along similar lines in individual countries, in the EU and in the wider international community. The main international body that has addressed the issue of food additive safety is the Joint FAO/WHO Expert Committee on Food Additives (JECFA). This Committee was set up in 1956 and over the years has drawn on expertise from around the world for its changing membership. Over the first 5 years of its existence, JECFA set out principles for the assessment of food additives, which have been collated and updated in subsequent years (10). The general principles of the JECFA approach have been widely adopted by other national and international bodies, including the SCF or, since May 2003 by EFSA.

Derivation of an acceptable daily intake

The JECFA approach is based on assessment of a usually extensive series of toxicological tests, identification of any critical toxic effects, their dose-response relationships, the doses at which they do not cause any adverse effects, and the setting of an Acceptable Daily Intake (ADI). The ADI is defined by JECFA as an estimate of "the amount of a food additive, expressed on a body weight basis, that can be ingested daily over a lifetime without appreciable health risk" (10). The ADI is derived by applying a safety or uncertainty factor to (usually) the lowest no-observed-adverse-effect level (NOAEL) in the toxicity studies. The safety factor most commonly used is 100, comprising a factor of 10 to take account of possible inter-species differences when extrapolating from animal experiments to humans and a further factor of 10 to take account of possible inter-individual differences between humans. The ADI is expressed as a range from 0 to an upper limit in mg/kg body weight. For some food additives, an "ADI not specified" is allocated. This is because a number of additives, in contrast to, say, pesticides or drugs, are of very low toxicity and no toxic effects are seen during animal testing when large amounts are given in the diet. In some instances, they may be the same as normal food ingredients (e.g. citric acid) or human metabolites (e.g. carbon dioxide, lactic acid). For some additives, EFSA may make their own decision in relation to setting maximum permitted levels for food additives irrespective of JECFA's opinions. For example, in the case of starch aluminium octenyl succinate, JECFA did not set any ADI levels as no toxicity data was available in 1997 and also as it was considered that the intake of aluminium from this source would be low and hence would not pose a safety concern (11). However in 2006, starch aluminium octenyl succinate became a newly permitted miscellaneous additive within the EU, up to a maximum level of 35 g/kg permitted in encapsulated vitamin preparations in food supplements.

On the other hand, there may be considerable evidence of safe human use of an additive from a country outside the EU. An example here would be in the case of Stevioside which is extracted and refined from *Stevia rebaudiana Bertoni* leaves and is used as a sweetener. It is permitted for use in certain countries including Japan, however it is not permitted on the market as a food or food ingredient in the EC or the US due to insufficient data on its safety (12).

Structurally related additives, with a common mechanism of action or effect, may be assigned a Group ADI. The summed intakes of all additives in a group should not exceed the figure for the Group ADI.

Toxicological tests required

The range of toxicological tests generally required for a proposed new food additive has been set out by various bodies, including JECFA (10), the SCF (13) and the US FDA (14). While these various guidelines differ in some aspects of detail, the core requirements are very similar. The SCF guidelines for submissions for food additive evaluations have been endorsed by EFSA (2nd meeting of AFC Panel, 9 July 2003) (13) and these guidelines replace those given by the SCF in 1980 on the toxicological tests generally required for additives.

Acute toxicity studies

Acute toxicity studies are not mandatory for the safety assessment of food additives although such data will often exist because of their necessity for occupational safety assessments in manufacturing. Hence, if such studies have been conducted for other purposes then they should be submitted. Similarly, studies on eye and skin irritation and skin sensitisation may also be available for occupational reasons but are of little use in the safety evaluation of food additives as consumed, hence they are not required. Skin sensitisation, for example, is not predictive of oral sensitisation, for which no validated animal models currently exist.

Absorption, distribution, metabolism and excretion

Studies on absorption, distribution, metabolism and excretion are usually conducted following single and short-term repeat dosing. These can greatly aid in the design of subsequent toxicity tests, indicate whether harmful metabolites may be produced, or whether the parent compound or its metabolites may accumulate in the body, and help in the interpretation of adverse findings in toxicity tests. Information on metabolism in humans is desirable. This occasionally becomes available for food additives after they have been approved and marketed. More such data could be safely generated prior to approval by the use of very small doses of radiolabelled compound, as is done in human volunteers for therapeutic drugs. Only when such comparative data are available can a definitive judgement be made on whether appropriate species have been selected for toxicity testing. *In vitro* studies can also give additional useful information.

Sub-chronic, repeat-dose toxicity studies

Sub-chronic studies in rodent and non-rodent species (usually 13-week studies in the rat and dog) for a period of at least 90 days are generally required. Ideally, exposure to the test compound should be via the oral route, usually given at fixed concentrations in the diet, but sometimes by gavage. For non-toxic substances, use of upper concentrations in the diet greater than 5% are not encouraged, since such concentrations tend to cause nutritional problems, which may then give rise to secondary toxicity. These studies yield important information on any food consumption and body weight changes, haematological changes, effects on blood and urine biochemistry which provide indications of damage to organs such as the liver and kidney, organ weight changes, and pathological changes in organs and tissues at the gross, macroscopic and microscopic levels.

Reproductive and developmental toxicity studies

Reproductive and developmental toxicity studies are also usually required. These generally comprise a multigeneration study in the rat and developmental toxicity studies in two species. Multigeneration studies assess any effects on male or female fertility, and the ability to maintain pregnancy, deliver offspring and maintain successful lactation; they also indicate any adverse effects on survival, growth and development of the offspring. Nowadays, such studies are likely to include not only assessment of the postnatal physical development of the offspring but also measures of motor and behavioural development. Such studies usually extend for either two or three generations so that the reproductive function of offspring themselves exposed to the test compound in utero can be assessed. This is currently regarded as very important, since critical aspects of the development of the reproductive system in rats occur in the late prenatal and early postnatal period – a developmental window in which there may be particular vulnerability to endocrine disrupter-induced effects.

In developmental toxicity studies (formerly known as teratology studies), the growth and development of the embryo and foetus are assessed, with emphasis on embryonic and foetal survival, foetal weight and the occurrence of any malformations. The dosing period for such studies was formerly during embryogenesis only (e.g. day 6–15 in the rat or 6–28 in the rabbit), but now it is recommended to continue dosing throughout embryogenesis and up to term (e.g. to day 21 in the rat). This is in order to cover important periods of brain and reproductive system development, which continue beyond the classical period of organogenesis for other systems. Dosing is normally via the diet or by gavage.

Genotoxicity studies

Genotoxicity studies assess the ability of a substance to interfere with DNA by induction of single gene mutations, chromosome aberrations, or other forms of DNA damage. Such effects on DNA are of significance because they indicate the potential for carcinogenic effects or induction of heritable mutations in germ cells. A battery of three *in vitro* genotoxicity tests are required (two for gene mutations and a chromosome aberration test), using bacterial systems such as *Salmonella typhimurium* and mammalian cells in culture from rodents or human lymphocytes. *In vivo* tests may also be required, especially if positive results are obtained in any *in vitro* studies, so that the genotoxic potential of the substance in the context of its *in vivo* metabolism and kinetics can be assessed. Substances that are genotoxic *in vitro* but not *in vivo* (e.g. because they are readily broken down into non-genotoxic compounds) are not generally regarded as hazardous to humans. For food additives, carcinogenicity studies (see below) will also normally be available to provide corroborating data for any genotoxic activity.

Chronic toxicity/carcinogenicity studies

Chronic toxicity/carcinogenicity studies in two species, usually rat and mouse, are also required for most food additives. Dosing commences when the animals are in the juvenile period of rapid growth, at about 6 weeks of age, and continues for most of the animal's lifetime (2 years or more for rats and mice). Dosing is almost invariably via the diet. The emphasis in these studies is on body weight, organ weight and pathological changes in tissues and organs. Examination of haematological and clinical chemistry parameters may also be included in satellite groups killed at intervals before the termination of the study at 2 years. Combined chronic toxicity/carcinogenicity studies are also considered acceptable.

Other studies

The core studies are designed to give clear information on the nature of any toxicity and NOAELs for most toxicological end points. However, for some less common aspects of toxicity, they are designed only to indicate a potential problem, and further studies may be required to elucidate these properly. Depending on the findings in these core tests, further special studies may sometimes be needed – for example, to clarify the mechanism of toxicity, in order to determine its relevance to humans, or to better define the (cellular, subcellular, biochemical) NOAEL. Similarly, if the core studies indicate, for example, that there may be effects on the immune, nervous or endocrine systems, further special studies designed to answer specific questions on these aspects may be required.

Test protocols and EC submissions

For the core tests, standard protocols are available, which have been developed and are widely accepted internationally. Studies conducted to OECD Guidelines (15) or EC Guidelines (16-18), the latter being essentially the same as OECD's, are acceptable for the testing of food additives for applications made to the EC. Further information on the presentation of applications to the EC for use of a new food additive has been published by the Commission (19).

Interpretation of toxicity tests

The toxicity of most food additives is generally low in comparison with that of other classes of chemical, such as pesticides, drugs and some industrial chemicals. The majority of effects observed in toxicity studies on food additives, usually confined to the higher levels of administration, are effects on body weight, with or without accompanying reductions in food consumption. Effects on the liver and kidney are also seen because these are the major organs of metabolism and elimination, so are often exposed to the highest concentrations of the additive and its metabolites. In reproductive and developmental toxicity studies, effects on the offspring, such as death and reductions in birth weight and postnatal growth, have to be assessed in light of whether the substance causes any maternal toxicity, since maternal toxicity can induce secondary effects on the offspring; this can often be a difficult judgement to make. In chronic toxicity/carcinogenicity studies, the maximum dose of a substance used should not cause undue mortality but should provide some evidence of toxicity (e.g. a reduction in body weight of up to 10% in comparison with controls). In this way, the maximum tolerated dose is given but the general toxicity should not interfere unduly with interpretation of the results with respect to carcinogenicity.

Relevance of effects observed for humans

In reviewing all the available toxicity studies, a judgement has to be made about which effects are adverse and which are not. For example, the feeding of large amounts of poorly absorbed materials, such as polyols, to rats is known to cause caecal enlargement, disturb calcium homeostasis, cause pelvic nephrocalcinosis and perhaps result in the development of adrenal phaeochromocytomas (20). However, this would not be taken as indicative of the same adverse effects occurring in humans if the additive were used in small amounts. On the other hand, the feeding of lower amounts of poorly absorbed bulk sweeteners, such as the polyols, can also cause an osmotic diarrhoea – an effect that is transient but which also occurs in humans, and this effect is taken into account in deciding in what foods such additives should be used and in setting maximum levels of use (20). There are also effects that occur in rodents that are not of significance for man, such as kidney damage and tumours via a mechanism involving a-2m-globulin – a protein formed only in male rat liver, which binds with certain hydrocarbons and accumulates in the kidney (21).

Effects may also be observed that are not regarded as being of toxicological significance, such as staining of tissues when high amounts of colours are fed to animals, or increases in liver weight and liver enzyme induction in response to metabolic overload when high amounts of some substances are fed. Similarly, sporadic but statistically significant changes in biochemical or haematological parameters, inevitable in any series of repeat-dosing studies, that are not accompanied by corroborating pathological changes may be disregarded.

Genotoxicity and carcinogenicity

The interpretation of genotoxicity and carcinogenicity studies is of special significance for food additives. Any substance that is genotoxic *in vivo* would not be regarded as acceptable for use as a food additive, since such effects may be without a threshold and thus could occur at very low daily exposures over a lifetime. Carcinogenicity bioassays often confirm the adverse consequences of genotoxic activity *in vivo*. If a substance is non-genotoxic *in vivo* but does show evidence of carcinogenicity in lifetime rodent studies (e.g. the sweetener sodium saccharin), it may still be acceptable as a food additive, provided a mechanism of toxicity and a threshold for its action can be identified (20). Such substances may act by inducing tissue damage and necrosis, resulting in enhanced cell division during repair, and tumour development. Provided a repeat-dose causing no tissue damage can be identified, an ADI may be set by applying an uncertainty factor to the NOAEL for tissue damage. For example, the antioxidant butylated hydroxyanisole (BHA) causes forestomach tumours in the rat when fed at 1 and 2% in the diet, via prolonged stimulation of the stomach epithelium causing hyperplasia. Hyperplasia, but not tumours, are also seen at 0.5% in the diet, but there is a NOAEL for hyperplasia of 0.125% in the diet.

Setting the ADI

In the setting of ADIs for food additives, it is often long-term, chronic toxicity studies or multigeneration studies that are critical in determining the lowest NOAELs and hence the ADI. This is due to the long periods of administration and the fact that they cover a number of critical periods in the lifetime.

To determine the ADI, a default safety or uncertainty factor of 100 is usually applied to the lowest NOAEL, unless other considerations intervene. If, for example, the critical study is one involving human subjects, then a reduced safety factor of perhaps 10 may be applied. Such is the case, for example, with the colour erythrosine. This affects the human and rat thyroid, ultimately causing tumours in the rat due to excessive production of thyroid stimulating hormone (TSH). Its mechanism of action is well understood and a no-effect level for increases in thyroid hormone levels in humans has been established and used, in conjunction with a 10-fold rather than a 100-fold safety factor to set the (low) ADI (23).

The majority of ADIs set nowadays by EFSA are full ADIs, but temporary ADIs are sometimes set by EFSA and by JECFA. A temporary ADI may be set when the data are sufficient to determine that no harm is likely to result from consumption of the additive over a short period of time, but that some further piece of information is needed to complete the database or to answer a specific question, in order to provide reassurance about lifetime exposure. A deadline is usually set for the submission of the required data. In such cases, an additional safety factor of 2 or more may be employed in setting the temporary ADI to take account of the residual additional uncertainty. If there are large gaps in the database, the AFC panel does not allocate an ADI, either temporary or full. Higher overall safety factors than 100 are also sometimes applied to take account of the severity or irreversibility of a critical effect (e.g. if teratogenicity or non-genotoxic carcinogenicity are the effects determining the ADI), the rationale being to err on the side of caution. An additional safety factor of 10 may also be applied if there is a minimal toxicity level apparent but no clear NOAEL from the toxicity studies.

Comparing Intakes with ADIs

The advantage of regulatory and advisory bodies setting ADIs for food additives is that they are universally applicable in different countries and to all sectors of the population. The one exception to this is in the case of food additives for infant formulae, for which EFSA considers it may be necessary to conduct specially designed additional tests and perhaps to set a different ADI (24). This is because standard toxicity testing protocols do not adequately model artificial feeding in the neonatal phase.

Methods for estimating food additive intake

To assess the health significance, if any, of intakes of food additives, the ADI can be compared against average and extreme consumption estimates in the population as a whole, or in particular subgroups of the population (e.g. sweetener intakes in diabetics). There are a large number of practical problems in estimating dietary intakes of food additives, and obtaining reliable estimates of average and extreme consumption amongst various sub-groups of the population with differing dietary habits is both time-consuming and expensive (25). It is therefore more usual for food additive intakes to be estimated initially using relatively crude approaches and for these to be further refined if necessary.

A very rough estimate of food additive intake on a national scale can be made by dividing the total weight of a food additive made annually, or the disappearance annually of a food additive into the food chain, by the number of individuals in the population as a whole. However, the annual per capita consumption figure generated by this means may be misleading in that it usually does not take account of imports and exports of food, and it assumes that all food sold is consumed. Overall, it is thought to considerably underestimate the actual exposure of individuals because it assumes consumption is even across the entire population, which is rarely the case. Such per capita estimates are rarely useful for providing reassurance that ADIs are not being exceeded. One way of reducing the likely underestimate is to assume that only 10% of the population are consumers. This method is being used by JECFA to estimate intakes of flavouring substances (26).

An initial screening method that has gained popularity in Europe is known as the (Danish) Budget Method (27). It relies on assumptions regarding physiological requirements for energy and liquid and on energy density of foods, instead of on detailed food consumption surveys. It assumes that all foods contributing to energy intake and all beverages contributing to liquid intake will contain the additive at the maximum permitted use levels. The resulting intake estimation is clearly an overestimate, but if such an overestimate is below the ADI for the additive concerned, no further refinement of the intake estimate is necessary.

A more refined method is to use surveys of food consumption that are representative of the population as a whole, and which provide information on average and extreme consumption of a wide range of foods. It is then assumed that all foods that may contain any particular additive do, and that they do so at the maximum level permitted or maximum level needed to achieve the desired technological effect. Combining these two types of information results in estimated intakes for average and extreme consumers that are thought to be conservative in that the assumptions made about food additive content are likely to overestimate actual intakes (28). If a highly refined estimate of intake is needed, detailed individual intake data can be gathered by means of food diary studies or duplicate diet studies. In food diary studies, estimates of intake may be made from the self-reported record of the individual's food consumption and information on the additive content of each food, obtained either from the manufacturer of the food or from knowledge of the maximum likely additive content of particular types of food. In duplicate diet studies, more accurate estimates of intake may be made by analysing exact replicates of the food eaten by an individual for the additive content. The latter method is particularly costly, and both of these methods are limited by the number of individuals that can be surveyed.

Estimates generated for food additive consumption are then compared with the ADI. Since most food additives are regularly ingested and not of any direct health benefit to the consumer, it is important to make this comparison not only for average consumers, but also for extreme consumers, to ensure that they are protected. Different regulatory authorities may use differing cut-off points to define "extreme" consumers. None uses the most extreme consumer since, within any population, there are likely to be one or two individuals with bizarre dietary habits, whose intakes are completely unrepresentative of intakes among the vast majority of the population. Extreme consumers are therefore usually defined as those having the 90th, 95th or 97.5th percentile of intake for the population as a whole.

Significance of exceeding the ADI

Provided that intakes for average and extreme consumers are within the ADI, it is reasonable to assume that there is very unlikely to be any risk to health. Even if intakes occasionally exceed the ADI, it is unlikely that any harm will result, since the ADI is based on a no-effect level and not on an effect level, to which a large safety factor has also been applied. Moreover, ADIs are often derived from long-term studies in which the additive is administered daily over a lifetime and prolonged administration at the toxic dose was required to elicit an effect.

If, however, intake figures indicate that the ADI may be regularly exceeded by certain sectors of the population, it may be necessary to reduce levels in foods or reduce the range of foods in which the additive is permitted for use. Since levels in foods are determined by what is needed to achieve the desired technological effect, the option to reduce levels in foods may not be available. It can, however, be considered when there are several additives within a class (e.g. colours, sweeteners, antioxidants) that perform the same function and which may be used in combination with each other.

Even when intake, by for example the 97.5th percentile of consumers, is within the ADI, it may be necessary to consider those individuals exceeding it to determine whether they represent a discrete population subgroup with atypical dietary behaviour (e.g. diabetics), which predisposes them to exceed the ADI. Action such as specific targeted advice or modification of particular products may then be necessary.

In cases where the ADI is determined by an irreversible effect that can occur following short-term rather than long-term exposure, particular care may be necessary to ensure that the ADI is not exceeded, even on an occasional basis (29). This would be the case, for example, if the ADI were determined by a developmental effect, since compounds affecting embryonic and foetal development can do so after exposure of only a matter of days. However, even in these cases, because large safety margins are usually built into the ADI for such serious, irreversible effects, it is likely that the ADI would need to be exceeded by some considerable margin for there to be any risk of harm to human health. A more comprehensive discussion of the significance of excursions of intake above the ADI is available (30).

Conclusions

The use of additives in foods traded within the EU is strictly controlled by legislation, which can be amended to include newly approved additives or (more rarely) to delete additives that are no longer approved. Food additives can be approved for inclusion in EU 'positive lists' only after full consideration of safety aspects by EFSA, which advises the Commission. The AFC Panel usually sets an ADI or, in the absence of an ADI, may stipulate other limitations on conditions of use. Once marketed, the monitoring of consumer intakes of food additives by the Members States enables checks to be carried out to ensure that Acceptable Daily Intakes are not regularly exceeded.

References

European Union legislation can be downloaded from the Europa (eur-lex) Web site at:
http://eur-lex.europa.eu/en/index.htm

References to the principal directives only are given below for References 2-5 and 7-9. When searching on eur-lex, the amendments to these directives are listed under the 'Simple Search' and 'consolidated text' facility.

1.   Completion of the Internal Market, Community Legislation on Foodstuffs. Office for Official Publications of the European Communities. Luxembourg, COM(85)603, 1985.

2.   Council Directive 89/107/EEC of 21 December 1988 on the approximation of the laws of the Member States concerning food additives authorized for use in foodstuffs intended for human consumption. Official Journal of the European Communities. L40, 11.2.89, 27-33.

3.   European Parliament and Council Directive 94/35/EC of 30 June 1994 on sweeteners for use in foodstuffs. Official Journal of the European Communities. L237, 10.9.94, 3-12.

4.   European Parliament and Council Directive 94/36/EC of 30 June 1994 on colours for use in foodstuffs. Official Journal of the European Communities. L237, 10.9.94, 13-29.

5.   European Parliament and Council Directive 95/2/EC of 20 February 1995 on food additives other than colours and sweeteners. Official Journal of the European Communities. L61, 18.3.95, 1-40.

6.   Wagstaffe, P.J. The assessment of food additive usage and consumption: the Commission perspective. Food Additives and Contaminants. 1996, 13(4), 397-403.

7.   Commission Directive 95/31/EC of 5 July 1995 laying down specific purity criteria concerning sweeteners for use in foodstuffs. Official Journal of the European Communities. L178, 28.7.95, 1.

8.   Commission Directive 95/45/EC of 26 July 1995 laying down specific purity criteria concerning colours. Official Journal of the European Communities. L226, 22.9.95, 1-45.

9.   Commission Directive 96/77/EC of 2 December 1996 laying down specific purity criteria on food additives other than colours and sweeteners. Official Journal of the European Communities. L339, 30.12.96, 1.

10.   Principles for the Safety Assessment of Food Additives and Contaminants in Food. Environmental Health Criteria 70. International Programme on Chemical Safety (IPCS) in cooperation with the Joint FAO/WHO Expert Committee on Food Additives (JECFA). World Health Organisation, Geneva, 1987.

11.   Scientific Committee on Food (1999). Opinion on starch aluminium octenyl succinate (SAOS) (expressed on 21 March 1997). Reports of the Scientific Committee on Food (Forty-third Series). CEC, Office for Official Publications of the European Communities, Luxembourg.

12   Scientific Committee on Food (1999) Opinion on stevioside as a sweetener (adopted on 17 June 1999).

13.   Scientific Committee for Food (2001) Guidance on submissions for food additive evaluations by the Scientific Committee on Food (opinion expressed on 11 July 2001).

14.   US FDA (1993). Draft Principles for the Safety Assessment of Direct Food Additives and Color Additives Used in Food. US Food and Drug Administration, Center for Food Safety and Applied Nutrition, Washington DC.

15.   OECD (1987). Guidelines for the Testing of Chemicals and subsequent revisions. Organisation for Economic Co-operation and Development, Paris.

16.   Commission Directive 84/449/EEC of 25 April 1984, adapting to technical progress for the sixth time, Council Directive 67/548/EEC on the approximation of the laws, regulations and administrative provisions relating to the classification, packaging and labelling of dangerous substances. Official Journal of the European Communities. L251, 19.9.84, 1.

17.   Commission Directive 87/432/EEC of 3 August 1987, adapting to technical progress for the eighth time, Council Directive 67/548/EEC on the approximation of the laws, regulations and administrative provisions relating to the classification, packaging and labelling of dangerous substances. Official Journal of the European Communities. L239, 21.8.87, 1.

18.   Commission Directive 92/69/EEC of 31 July 1992, adapting to technical progress for the seventeenth time, Council Directive 67/548/EEC on the approximation of the laws, regulations and administrative provisions relating to the classification, packaging and labelling of dangerous substances. Official Journal of the European Communities. L383, 29.12.92, 1.

19.   Commission of the European Communities (1989). Presentation of an Application for Assessment of a Food Additive Prior to its Authorization. Office for Official Publications of the European Communities, Luxembourg. CB-57-89-370-EN-C.

20.   Scientific Committee for Food (1985). Report of the Scientific Committee for Food on Sweeteners (opinion expressed on 14 September 1984). Reports of the Scientific Committee for Food (Sixteenth Series). CEC, Office for Official Publications of the European Communities, Luxembourg.

21.   National Research Council (1996). Carcinogens and Anticarcinogens in the Human Diet. Washington DC, National Academy Press.

22.   Scientific Committee for Food (1989). Report of the Scientific Committee for Food on Antioxidants (opinion expressed 11 December 1987). Reports of the Scientific Committee for Food (Twenty-second Series). CEC, Office for Official Publications of the European Communities, Luxembourg.

23.   Scientific Committee for Food (1989). Report of the Scientific Committee for Food on Colouring Matters (opinion expressed 10 December 1987). Reports of the Scientific Committee for Food (Twenty-first Series). CEC, Office for Official Publications of the European Communities, Luxembourg.

24.   Scientific Committee on Food (1998). Opinion on the applicability of the ADI (Acceptable Daily Intake) for food additives to infants. Minutes of the 113th meeting of the Scientific Committee on Food held on 16-17 September 1998. Annex I to Doc XXIV/2210/98. CEC, DGXXIV, Brussels.

25.   ILSI Europe Workshop on Food Additive Intake: Scientific Assessment of the Regulatory Requirements in Europe, 29-30 March 1995, Brussels Summary Report. Food Additives and Contaminants. 1996, 13(4), 385-95.

26. Safety Evaluation of Certain Additives and Contaminants. Fifty-first report of the Joint FAO/WHO Expert Committee on Food Additives. WHO Food Additives Series 42. 1999.

27. An Evaluation of the Budget Method for Screening Food Additive Intake. Summary Report of an ILSI Europe Food Chemical Intake Task Force. ILSI Europe, Brussels, Belgium, April 1997.

28. Gibney M.J., Lambe J. Estimation of food additive intake: methodology overview. Food Additives and Contaminants. 1996, 13(4), 405-10.

29. Rubery E.D., Barlow S.M., Steadman J.H. Criteria for setting quantitative estimates of acceptable intakes of chemicals in food in the UK. Food Additives and Contaminants. 1990, 7 (3), 287-302.

30. ILSI Europe Workshop on the Significance of Excursions of Intake Above the Acceptable Daily Intake (ADI). Eds Barlow S., Pascal G., Larsen J.C., Richold M. Regulatory Toxicology and Pharmacology. 1999, 30 (No 2, Part 2).

Further Reading

Commission Decision 2000/196/EC 22 February 2000 refusing the placing on the market of *Stevia rebaudiana Bertoni*.

Directive 2006/52/EC of the European Parliament and of the Council amending Directive 95/2/EC on food additives other than colours and sweeteners and Directive 94/35/EC on sweeteners for use in foodstuffs.

4. LEGISLATION FOR FOOD ADDITIVES OUTSIDE EUROPE

Introduction

A major problem for the manufacturers of food products over the years has been that of trying to find ingredient and additive specifications that will enable a particular food product of interest to be sold in more than one country. The world is becoming smaller in terms of exporting food products, with many companies looking to sell their products in overseas markets other than Europe. However, additives legislation can differ significantly from country to country, both in how it is structured and in details such as the acceptability of named additives in individual foods. It is true that developments in key trading blocs such as the EU have influenced additives legislation in other parts of the world. The differences in how food additives are regulated are still a major concern to both ingredients and additives suppliers and also to manufacturers of the final food product.

Two factors that strongly influence the use of particular food additives are technological need and safety. It is necessary to ensure that levels of particular additives do not increase above acceptable safety limits for those people who, by nature of their diet, may consume high amounts of a particular additive, for example by the consumption of significant amounts of a sweetener through consumption of large quantities of soft drinks. The establishment of Acceptable Daily Intakes (ADIs) for additives is referred to throughout the chapter.

Labelling issues

Another factor to consider is labelling. There was a trend some years ago to replace chemical names on a label with equivalent numbers – for example, E numbers. The pendulum has now swung the other way, with manufacturers often preferring to use additive names wherever possible (taking account of the complexity of the chemical name and the limited room they may have on the label) and avoiding the use of numbers. Generally, today, a mixture of both is used. Consumers are considered to have a right to make an informed choice about the foods they consume, and most additives are generally required to be declared on the label as part of the ingredients list. However, those that are present only as a result of carry-over from inclusion in another ingredient, with no technological effect in the final food, or those used as processing aids, generally do not have to be declared unless they, or the raw materials from which they are derived, are known to cause allergies or intolerances. Exact rules for exemptions from additives declarations vary from country to country, but generally the same principles are followed. Compounds added for their nutritional benefit are not generally considered as food additives, but there are exceptions to regulatory provisions, and this is the case in Japan.

An additive used in a food for a technological purpose should not be declared as a nutrient, but with the category name of the function to which it relates. For example, if ascorbic acid is being added as an antioxidant, it should be declared as an antioxidant and not as vitamin C.

Purity criteria, i.e. the specification to which additives must be manufactured, are established in the USA (under Food Chemicals Codex), as part of EU legislation, at Codex level and sometimes by individual countries.

In this chapter, developments in food additives at Codex level are assessed and key aspects of international food legislation in major export nations outside Europe, such as the USA, Japan, Canada, Australia, major Latin American countries and other Far East markets are discussed.

Codex Alimentarius

The Food and Agriculture Organization (FAO) of the United Nations, and the World Health Organization (WHO) established the Codex Alimentarius Commission in 1963 after recognising a need for international industry standards for the food industry worldwide, in order to protect the health of consumers and to make international trade in food easier. Any Member Nation and Associate Member of the FAO and WHO that is interested in international food standards may become a member of the Commission, after notifying the Director-General of FAO or WHO. Non-members of the Commission with special interest in its work may attend Commission sessions, or sessions of its subsidiaries and ad hoc meetings as observers, provided they request to do so prior to the meeting. The EU is also a Member of Codex, even though the individual Member States are also Members. Nations that are not members, but belong to the United Nations, may be invited, if they request, as observers (1).

The aim of Codex is to prepare international voluntary standards, recommendations and guidelines in order to protect public health, ensure fair trading and promote harmonisation. The standards are developed by consensus and on the best available scientific and technical advice. A uniform procedure has been established for the development of Codex standards, known as the 'step' procedure. There are eight steps in the procedure; at step 8, a draft is submitted to the Codex Alimentarius Commission, plus any proposals for amendment at that stage, with a view to adoption as a Codex standard. An accelerated procedure can be followed, which is completed in 5 steps. Once published, Codex standards are sent to governments for acceptance and to international organisations with competence in the subject in question as named by the member governments. A list is then published of the countries that will allow products conforming to the standard in question to be freely distributed. In some cases, specific deviations may also be accepted and published, and are notified to the Commission as possible amendments. These publications together form the Codex Alimentarius. The Codex Alimentarius is a compilation of texts concerning the following, of which additives is just one part:

- Food standards
- Hygiene and technological practice
- Pesticide evaluations
- Maximum residue limits (MRLs) for pesticides
- Guidelines for contaminants
- Food additives evaluations
- Veterinary drugs evaluations

In many cases, recommended uses for additives are contained within individual Codex food standards. Much of the work in producing individual standards is handled by specific sub-committees, with final endorsement by the Codex Commission itself. The relevant committee for food additives is the Codex Committee on Food Additives and Contaminants (CCFAC) which was divided into two committees in 2006 - the Codex Committee on Food Additives (CCFA) and the Codex Committee on Contaminants in Foods (CCCF). The trend these days continues to be away from so-called 'product-specific' legislation to horizontal provisions aimed at all food types; this and the increasingly important role of Codex standards as reference texts in trade disputes under the World Trade Organisation agreements made it appropriate to establish a General Standard on Food Additives (GSFA). This follows the EU pattern of having additives and their food uses detailed together and separately from individual product standards.

Codex General Standard on Food Additives

The Codex General Standard for Food Additives was adopted as a Codex standard in 1995 (192-1995). Since then, there have been several amendments to this standard, with the latest being adopted in 2007. Only the additives that have been evaluated as safe by the Joint FAO/WHO Expert Committee on Food Additives (JECFA) and have been assigned an ADI and an International Numbering System (INS) designation by Codex are included in this standard. Only those additives included in this standard are suitable for use in foodstuffs. The standard gives details of conditions under which permitted food additives may be used, whether or not they have been previously standardised by Codex (2). The food additive provisions of Codex commodity standards are considered and superseded by the provisions of this general standard. The aim of establishing permitted levels of use of additives in various food groups is to ensure that the intake of additives does not exceed the acceptable daily intake. According to this standard, a food additive is defined as follows:

'any substance not normally consumed as a food by itself and not normally used as a typical ingredient of the food, whether or not it has nutritive value, the intentional addition of which to food for a technological (including organoleptic) purpose in the manufacture, processing, preparation, treatment, packing, packaging, transport or holding of such food results, or may be reasonably expected to result (directly or indirectly), in it or its by-products becoming a component of or otherwise affecting the characteristics of such foods. The term does not include 'contaminants' or substances added to food for maintaining or improving nutritional qualities.'

According to the preamble of the General Standard, the use of food additives is justified only when such use

- has an advantage for,
- does not present a hazard to the health of, and
- does not mislead,

the consumer and serves one or more of the purposes and needs established below and only when these cannot be achieved by other means that are economically and technologically practicable.

a. To preserve the nutritional quality of the food; an intentional reduction in nutritional quality of a food would be justified in (b) below and in other circumstances where the food does not constitute a significant item in a normal diet;

b. to provide necessary ingredients or constituents for foods manufactured for groups of consumers having special dietary needs;

c. to enhance the keeping quality or stability of a food or to improve its organoleptic properties, provided this does not change the nature, substance or quality of the food so as to deceive the consumer;

d. to provide aids in the manufacture, processing, preparation, treatment, packaging, transport or storage of food, provided the additive is not used to disguise the effects of use of faulty raw materials or undesirable (including unhygienic) practices or techniques during the course of any of these activities.

The food category system is hierarchical; when an additive is permitted in a general category, it is also permitted in all its subcategories, unless otherwise stated. It is based on food category descriptors as marketed and is also in compliance with the carry-over principle. An additive may be acceptable in a final food provided it is permitted in one of the component ingredients or raw materials according to this Standard and its amount in these does not exceed the maximum permitted, and provided that the amount of that ingredient present in the final food will not be higher than would be by the use of the ingredients under proper technological conditions or manufacturing practices. It is used to simplify the reporting of food additive uses for the development of this Standard.

The food additives listed in the Standard Preamble were grouped into 23 major functional classes. The draft revision of the Codex class names and the International Numbering System (INS) of July 2007 includes 27 classes of Food Additives (3). The INS system has been developed for the purpose of identifying food additives in ingredients lists as an alternative to the declaration of the specific name, which can be lengthy and complex. The INS does not imply toxicological approval by Codex but is purely a means of identifying additives on a worldwide basis. There is also a table of food categories excluded from the general conditions for food additives.

As the INS list is primarily for identification purposes, it is an open list subject to the removal of existing additives or the inclusion of new ones on an ongoing basis. The CCFA will maintain an ongoing review, in conjunction with the Codex Committee on Food Labelling, of the functional class titles specified for use in food labelling. Member governments and interested organisations can make proposals to CCFA on an ongoing basis regarding:

*   additional additives for which an international identification number can be justified;
*   additional functional class titles for use in food labelling in conjunction with the INS;
*   the deletion of food additives or class titles.

The General Standard is developed on an additive-by-additive basis rather than by functional class. This Standard contains a detailed list of additives permitted for use under specified conditions in certain food categories or individual food items – for example, Class III and Class IV caramels as colours in named foods, ferric ammonium citrate as an anticaking agent in named foods, and polydimethylsiloxane as an anticaking or antifoaming agent in named foods (Table 1). In the Standard there is also a list of additives permitted for food use in general, unless otherwise specified, in accordance with Good Manufacturing Practice (GMP) (Table 3).

According to GMP:

- *The quantity of additive added to food shall be limited to the lowest possible level necessary to accomplish its desired effect;*
- *The quantity of the additive that becomes a component of food as a result of its use in the manufacturing, processing or packaging of a food and that is not intended to accomplish any physical or other technical effect in the food itself, is reduced to the extent reasonably possible;*
- *The additive is prepared and handled in the same way as a food ingredient.*

This list includes a number of additives across a range of technical functions – for example, guar gum, beet red, papain, polydextrose, sodium ascorbate, xylitol and lecithin. These additives are not specified by function. However, although generally permitted in accordance with GMP, it is recognised that it is not in line with GMP to use additives in a number of unprocessed foods or basic foods without significantly changing the nature of the product. The use of generally permitted additives is restricted in products such as milk and buttermilk; fresh, surface-treated or peeled or cut fruit and vegetables; liquid and frozen egg products; honey; fats and oils that are essentially free of water; coffee and tea infusions; infant formulae and weaning foods; and dried pastas and noodles.

Overall the Standard comprises of three main components:
- A) Preamble

- B) Annexes
  - o Annex A: Guideline for considering maximum use levels for additives with numerical JECFA ADIs
  - o Annex B: A listing of the food category system used to develop and organise Tables 1, 2 and 3 of the standard. Descriptors for each food category and sub-category are also given.
  - o Annex C: A cross reference list of the food category system and Codex commodity standards.**

- • C) Food Additive Provisions
- o Table 1 specifies, for each food additive or food additive group (in alphabetical order) with a numerical JECFA ADI, the food categories (or foods) in which the additive is recognised for use, the maximum use levels for each food or food category, and its technological function. Table 1 also includes the uses of those additives with non-numerical ADIs for which a maximum use level is specified.
- o Table 2 contains the same information as Table 1, but the information is arranged by food category number.
- o Table 3 lists additives with Not Specified or Not Limited JECFA ADIs that are acceptable for use in foods in general when used at *quantum satis* levels and in accordance with the principles of Good Manufacturing Practice described in Section 3.3 of the preamble. The Annex to Table 3 lists food categories and individual food items excluded from the general conditions of Table 3. The provisions in Tables 1 and 2 govern the use of additives in the food categories listed in the Annex to Table 3

**At time of publication a revision of the Food Category System (FCS) of the Codex General Standard for Food Additives was underway.

JECFA

Key to food additives usage is assessment of their safety when consumed as part of manufactured products (4). It is the role of the Joint FAO/WHO Expert Committee on Food Additives (JECFA) to evaluate food additives in terms of their toxicology and to prepare specifications for each additive, including, where necessary, an ADI. Independent expert scientific advice is provided at an international level. Additives are only one interest of JECFA; it also deals with contaminants, veterinary residues and naturally occurring toxicants.

An ADI in this context is an estimate, by the Committee, of the amount of a food additive, expressed on a body weight basis, that can be taken daily in the diet, over a lifetime, without appreciable health risk. The weight of a standard man is taken as 60 kg. Generally, ADIs are expressed in mg/kg body weight, in a range from 0 to an upper limit, which is considered to be the zone of acceptability of the substance. The acceptable level established is an upper limit and the Committee encourages the lowest levels of use that are technologically feasible.

The ADI may be qualified at present by a number of terms:

Not specified – If an ADI is not specified, then, on the basis of available chemical, biochemical and toxicological data, the total daily intake of the substance arising from use at levels necessary to achieve the desired effect and from its acceptable background in food, does not, in the Committee's opinion, represent a hazard to health. An ADI is not, therefore, deemed necessary. An additive with a non-specified ADI must be used in accordance with Good Manufacturing Practice (GMP).

Not limited – a term no longer used by JECFA with the same meaning as not specified.

Conditional – a term no longer used by JECFA to signify a range above the unconditional ADI, which may signify the acceptable intake when special circumstances or special groups of population are considered.

Temporary – A temporary ADI may be allocated when insufficient data are available to establish the safety of a substance and it is considered necessary for the remaining information to be supplied within a stated period of time. A higher-than-normal safety factor is then used and a deadline to resolve the safety issue.

Not allocated – This term is used when an ADI is not established for one of the following reasons:

- Insufficient safety information is available.
- No information is available on its food use.
- Specifications for identity and purity have not been produced.

Group ADI – Additives that are closely related chemically and toxicologically can be grouped together for the purposes of evaluation, for example polyoxyethylene sorbitan esters, modified celluloses and phenolic antioxidants. The ADI is expected to cover all the members of the group that may be included in the diet.

TABLE 4.1
ADIs of some commonly used sweeteners

Additive	ADI (mg/kg body weight)
Calcium cyclamate	0-11 (expressed as cyclamic acid). Group ADI for calcium and sodium salts
Aspartame	0-40 for aspartame and 0-7.5 for diketopiperazine
Acesulfame	0-15
Saccharin	0-5 Group ADI for saccharin and Ca, K, Na salts, singly or in combination

Specifications

Specifications for the identity and purity of food additives developed by the Committee have three purposes;

- to ensure that the additive has been biologically tested;
- to ensure that the substance is of the quality required for the safe use in food;
- to reflect and encourage Good Manufacturing Practice.

Specifications include additive synonyms, definition, assay, description, functional uses, tests of identity and impurities and assay of major components. The specifications are periodically reviewed, owing to changes in patterns of additive use, in raw materials and in methods of manufacture. Specifications may be full or tentative; 'tentative' is used only in cases where data on the purity and identity of the substance were required. If no qualification of 'tentative' is made, then the implication is that the toxicological data and data for specifications are adequate.

Additives in accordance with the General Standard should be of appropriate food-grade quality and conform with the specification recommended by Codex, i.e. those set by JECFA, or, in the absence of such specifications, with appropriate specifications developed by responsible national or international bodies. Food-grade quality is achieved by compliance with the specification as a whole and not merely with individual criteria in terms of safety. JECFA has recently adopted principles concerning the governing of toxicological evaluations and specifications, which would be used to promote consistency in the decision-making process. Such principles have been requested for review and comment by members of government delegations to CCFA. JECFA meetings are usually held annually, as are CCFA meetings; the list of substances scheduled to be evaluated and requests for data are normally issued in advance of these meetings.

Food Additives Legislation in Other Countries

USA

Food additives legislation in the USA has evolved in a unique manner and it is appropriate to consider how it differs from that applying elsewhere.

Framework

The Federal Food, Drug and Cosmetic Act (FFDCA) lays down the framework for food safety at a Federal level in the USA. This includes the definitions and principles on the use of food additives. The provisions of the Act are enforced by the Food and Drug Administration (FDA) through more detailed regulations laid down in Title 21 of the Code of Federal Regulations (21 CFR). The FDA is responsible for all food products including dietary supplements, and all food additives; however, other agencies have specific responsibilities for eggs, meat, poultry and alcoholic beverages. Despite this, the overall framework for the regulation of food additives is the same. The US Department of Agriculture (USDA) is responsible for meat and poultry products, and provisions on permitted additives in these products are laid down in Title 9 of the Code of Federal Regulations (9 CFR). The Bureau of Alcohol, Tobacco and Firearms (ATF) is responsible for alcoholic beverages, and permitted additives for these products are laid down in Title 27 of the Code of Federal Regulations (27 CFR).

The Food Additives Amendment was enacted to the FFDCA in 1958 and forms the present basis of the regulation of food additives. This Amendment defined the terms 'food additive' and 'unsafe food additive' and established a pre-market approval process for food additives. The adulteration provisions were amended to deem any food that contains any substance not regulated by these provisions as unsafe.

In accordance with the Act, the term 'food additive' means any substance of which the intended use results, or may reasonably be expected to result, directly or indirectly, in its becoming a component of any food, or otherwise affecting the characteristics of the food. This includes any substance intended for use in manufacturing, processing, treating or holding food, and any source of radiation used.

In enacting the Amendment, Congress recognised that many substances that were intentionally added to food would not require a formal pre-market review by the FDA to assure their safety. This was either because their safety had already been established by a long history of use in food or by virtue of the nature of the substances and the information generally available to scientists about the substances. As a result, the definition of a food additive excludes substances that are generally recognised as safe (GRAS), under the conditions of their intended use, among experts qualified by scientific training and experience to evaluate their safety. The view that a substance is GRAS may be based on scientific procedures or, for substances used in food prior to January 1958, on experience derived from its common use in food.

Colour additives are also excluded from the definition of a food additive and there are separate provisions on these laid down in the Act.

Therefore, substances intended for use in the manufacture of foodstuffs for human consumption are classified into three categories: food additives, prior-sanctioned food ingredients and substances generally recognised as safe (GRAS). In addition, there are separate provisions on colour additives. These will be described in turn.

TABLE 4.II
Additives classification in the USA

Class	Example
Food additive	Saccharin
Prior-sanctioned food additive	Sodium nitrate
GRAS substance	Sorbitol

21 CFR Parts 170–189 lay down regulations on these food additives and GRAS substances in detail, including the procedures for their approval, labelling requirements, specifications and purity criteria. In order to clarify the provisions of their use, 43 general food categories and 32 physical or technical functions have been established.

Food additives

A food additive may be direct, secondary direct or indirect, depending on how it is used.

Direct food additives are divided into eight categories: food preservatives; coatings, films and related substances; special dietary and nutritional additives; anticaking agents; flavouring agents and related substances;

gums, chewing-gum bases and related substances; other specific usage additives; and multipurpose additives.

Secondary direct food additives are components used in ingredients of processed foods that may become additives in the final food. These are divided into four categories: polymer substances and polymer adjuvants for food treatment; enzyme preparations and microorganisms; solvents, lubricants, release agents and related substances; and specific usage additives.

Some direct food additives are listed separately as those permitted on an interim basis pending additional study. This is where new information raises substantial questions about the safety or functionality of a substance, although there is reasonable assurance that no harm to public health will result from continued use of the substance while further study is carried out. Substances listed in this section include saccharin and its salts, and brominated vegetable oil.

Indirect food additives are materials that may become part of a food as a component of packaging material, adhesives, food-processing equipment, surfaces and containers used for food handling, and certain production aids and sanitisers. Substances used in food-contact articles that may be expected to migrate into food at negligible levels may be exempted from food additive status by petition. However, this would not be necessary if the substance was considered GRAS, or GRAS for use in food packaging, or was a substance that had prior-sanction approval (see below).

Food additives are approved for use by a petition procedure. Details of this are laid down in the legislation. The FFDCA requires that a regulation regarding the use of a food additive is issued 90 days after the petition is filed unless the time is extended to 180 days. However, in practice, food additive petitions can take much longer than this as the clock stops every time new information is required. The FDA has an expedited review procedure for certain food additive petitions that are expected to have a significant impact on food safety. These petitions are placed at the beginning of appropriate review queues and apply to additives that are intended to decrease the incidence of foodborne illness through antimicrobial action against human pathogens or their toxins in or on food (5).

Guidance for industry regarding the petitioning procedure and details of pending food additive and colour additive petitions are available via the FDA Center for Food Safety and Applied Nutrition (CFSAN) Internet site (6).

Prior-sanctioned ingredients

Prior-sanctioned ingredients listed in the legislation are substances that received official approval for their use in food by the FDA or the USDA prior to the Food

Additives Amendment in 1958. All the substances listed in this section are components of food-packaging materials, with the exception of sodium nitrate and potassium nitrate.

Substances that are Generally Recognised as Safe (GRAS)

21 CFR 182 lists some substances that were used in food prior to 1958 without known detrimental effects, whose regulatory status was clarified by the FDA. The regulations specifically state that it is impractical to list all substances that are GRAS for their intended use, such as common food ingredients and monosodium glutamate.

21 CFR 184 details substances that have been affirmed as GRAS for particular purposes after review by the FDA. However, this list does not contain all GRAS substances as the responsibility for proof of safety lies with the additive manufacturer. The list of substances that have been affirmed as GRAS includes many substances that are widely seen as food additives, such as locust bean gum, ammonium sulphate and adipic acid; in contrast, many substances that are typical food ingredients are also listed, such as garlic, sucrose and malt syrup.

It is the use of a substance, rather than the substance itself, that is eligible for GRAS exemption. This means that an affirmation may be granted for a particular use without taking into account other uses that may also be GRAS.

A determination that a particular use of a substance is GRAS requires both technical evidence of its safety and a basis to conclude that this evidence is generally known and accepted. In contrast, a determination that a food additive is safe requires only technical evidence of safety. Therefore, a GRAS substance is distinguished from a food additive on the basis of the common knowledge about the safety of the substance for its intended use rather than on the basis of what the substance is or the types of data and information that are necessary to establish its safety. In addition, the FDA has pointed out that the existence of a severe conflict among experts regarding the safety of the use of a substance precludes a finding of general recognition (7).

It is important to note that the GRAS exemption applies to the premarket approval requirements for food additives only and there is no corresponding exemption to the premarket approval requirements for colour additives. In other words, no colour additive in the USA has GRAS status.

Manufacturers may petition the FDA to affirm that a substance is GRAS under certain conditions of use. This petition process provides a mechanism for official recognition of GRAS determinations. GRAS petitions are in the public domain, whereas additives petitions are not, and unlike additive petitions, there is no time limit laid down in the legislation for the petition.

The FDA has recognised that the petition process takes up significant amounts of its resources and, on 17 April 1997, it issued a proposal to replace this procedure and to clarify when the use of a substance is GRAS (7). The proposed rule would replace the GRAS affirmation process with a notification procedure. This means that any person could notify the FDA of a determination that the use of a particular substance is GRAS. Within 90 days of receipt of the notice, the FDA would respond to the notifier in writing and advise the notifier that the Agency had identified a problem with the notice or otherwise. However, the FDA would not conduct its own detailed evaluation of the data that the notifier relies on to support a determination that a use of a substance is GRAS; nor would it affirm that the substance was GRAS for its intended use. In the interim between this proposal and any final rule, the FDA has encouraged use of this procedure. At the time of printing, a Final Rule has not been issued, and the GRAS notification program is still active "in the interim".

Since the issuing of the proposed rule, the FDA has received and responded to over 200 of such GRAS notices for a variety of substances, and has published an inventory of these notices and the Agency's response. This is available on the Internet and is updated on an approximately monthly basis (8).

Examples of current GRAS notifications on the Net

Among the notifications currently available for viewing on the Web site are those for a number of enzymes (including transglutaminase and alpha-amylase from named sources), calcified seaweed for use in yoghurts and biscuits, extract of bitter cola seed, menhaden oil, tasteless smoke, hydrogen peroxide as a processing aid for use in the storage of onions, and polyglycerol polyricinoleate (PGPR) as emulsifier (this to max. 0.3% in chocolate). These illustrate the range of compounds for which GRAS notification can be given; some of the compounds are more 'traditional' additives and others would be more likely to be classified as ingredients in Europe, albeit with specific approval required as novel ingredients in some cases.

Example of additive approval – olestra

Another example of differences in classification is that of olestra. Under European law, approval for olestra is now being considered by means of the EC novel foods Regulation, Regulation (EC) No. 258/97 rather than additives legislation. This is due to additives legislation in Europe being defined by the function of the compound in question. The mode of action of olestra takes the compound outside the scope of the additives legislation.

Title 21 172.867 of the Code of Federal Regulations defines olestra as a mixture of octa-, hepta- and hexa-esters of sucrose with fatty acids derived from edible fats and oils or fatty acid sources that are generally recognised as safe or approved for use as food ingredients. Various specifications are laid down. Olestra is authorised for use in place of fats and oils in prepackaged ready-to-eat savoury snacks (i.e. salty or piquant but not sweet) and prepackaged, unpopped popcorn kernels that are ready-to-heat. In such foods, the additive may be used in place of fats and oils for frying or baking, in dough conditioners, in sprays, in filling ingredients or in flavours. To compensate for any interference with absorption of fat-soluble vitamins, alpha-tocopherol, retinol equivalent, vitamin D and vitamin K have to be added in quantities as specified. The label of a food containing olestra must carry the following information:

The added vitamins must be included in the list of ingredients, but are not considered in determining nutrient content for nutrition labelling or any nutrition claims, express or implied. An asterisk shall follow vitamins A, D, E, and K in the listing of ingredients, appearing as a superscript following each vitamin. Immediately following the ingredient list an asterisk and statement, "Dietarily insignificant" shall appear prominently and conspicuously.

In order to be consistent with its obligation to monitor the safety of all food additives, the FDA stated it would review and evaluate all data and information bearing on the safety of olestra received by the Agency after the Regulation came into force. Such data, information and evaluation would then be presented to the Agency's Food Advisory Committee within 30 months of the effective date of the Regulation. This occurred at an open public meeting, held June 15-17, 1998, in which new data and information concerning olestra, obtained since the 1996 approval were presented. The complete set of transcripts of the June 15-17, 1998, FAC meeting is publicly available through the FDA's Division of Dockets Management and through the FDA's Internet site. One outcome of this meeting, in conjunction with a petition from a food manufacturer, was that the FDA amended 172.867 to remove a requirement for a warning statement concerning the possible undesirable effects of olestra on the gastro-intestinal system, and its effect on the absorption of some vitamins and other nutrients.

Flavours

Flavours may be either food additives or GRAS. Some of these are specifically laid down in the food additives and GRAS provisions. In addition, the US Flavor and Extract Manufacturers' Association (FEMA) also publishes its own list of flavouring agents that it considers GRAS for food use – the 'FEMA/GRAS list'. While the legislation does not specifically state that all flavouring agents appearing on the FEMA/GRAS list are approved for food use, the FEMA expert panel is widely recognised as complying with the GRAS exemption, which requires the safety of a substance to be evaluated by experts that are experienced and trained in evaluating the safety of food substances.

Colours

Legislation for food additives used as colours in the USA is especially strict. All substances that are deliberately used for their colouring effect are colour additives. The FFDCA allows the use of approved colours only for this purpose. Provisions on specifically permitted colours are laid down in 21 CFR Parts 73 and 74. All colour additives are classed as 'artificial'. They are divided into those that are exempt from certification and those that are subject to certification. All synthetic organic colours are subject to certification. This means that a sample from each batch is sent to the FDA, which determines the purity of the batch by laboratory analysis. Other colour additives are listed as those that are exempt from certification.

Colours that are subject to certification include: FD&C blue no. 1 (brilliant blue FCF), FD&C blue no. 2 (indigo carmine), FD&C green no. 3 (fast green FCF), FD&C red no. 3 (erythrosine), FD&C red no. 40 (allura red), FD&C yellow no. 5 (tartrazine) and FD&C yellow no. 6 (sunset yellow FCF). Other colours are listed provisionally and await re-evaluation. This includes the lakes of these colours (except for the lake of FD&C red no. 3).

Colours that are not subject to certification include: annatto extract, dehydrated beets, canthaxanthin, caramel, β-apo-8'-carotenal, β-carotene, cochineal extract, carmine, toasted partially defatted cooked cottonseed flour, grape colour extract, fruit juice, vegetable juice, carrot oil, paprika, paprika oleoresin, riboflavin, saffron, titanium dioxide, turmeric and turmeric oleoresin. Other substances are listed but these are restricted to specific uses only, such as animal feeds.

As for food additives, approval for colour additives is by petition to the FDA. A specific procedure for colour additive petitions is laid down in the legislation.

It is important to note that foodstuffs for which a standard is laid down may only contain colour additives if specifically permitted by that standard. Non-standardised foodstuffs, in general, may be coloured with permitted colour additives. However, the use of colour additives is not permitted if it conceals damage or inferiority, or if it makes the product appear better or of greater value than it is.

Labelling

Specific provisions on the labelling of some food additives and GRAS substances are laid down. Labelling provisions lay down general provisions on the labelling of colour additives and flavours. In contrast to the rules in many other countries, most additive functions do not need to be declared in a list of ingredients; however, the function of all chemical preservatives, leavening agents and firming agents must be stated. Warning statements are also required if specific additives are used. A warning statement for products containing saccharin has been laid down in the Act. The FDA has established other warnings for some additives, such as aspartame and sorbitol. For sorbitol, the label must declare 'excess consumption may have a laxative effect' if an amount of more than 50 g sorbitol is consumed on a daily basis.

Canada

It could be thought that Canada might base its additives legislation on that of its close neighbour, the USA, particularly with the introduction of the North American Free Trade Agreement (NAFTA). However, this is not the case. Currently, Canadian additives legislation is classified by food additive, with groups of additives being classed together by function (e.g. different classes of preservatives, colours, stabilisers, etc.). Individual additives may be listed as permitted in named foods or in unstandardised foods, or sometimes in both. The key to current additives usage is whether or not a food has a specific food standard.

Japan

Food additives legislation differs again in Japan. Instead of having just one list of permitted additives, as is often the case, there are two in Japan. The List of Designated Food Additives contains additives mainly produced by a chemical reaction (other than a degradation process) (9); designated food additives therefore include nature-identical additives. Some additives have specified restrictive conditions of use; others are generally permitted for food use, subject to any requirements being specified in compositional standards. A list of synthetic flavourings is included. The second list is the List of Existing Food Additives, which is a list of those natural food additives currently being used in Japan. Most of the natural additives on this list come with their own definition, which is often very precise. Care has to be taken when considering a synthetic equivalent of a natural compound, for example, β-carotene; the natural form may be regulated by the List of Existing Food Additives but the synthetic form by the List of Designated Food Additives.

The range of compounds included in the List of Existing Food Additives varies from those commonly recognised as food additives in other countries to compounds not generally regulated elsewhere (9). Examples of the former category include gum arabic (defined as a substance composed mainly of polysaccharides obtained from the secretion of acacia trees), chlorophyll, smoke flavourings, pectin and beet red. Examples of compounds included in the latter category include grapefruit seed extract (defined as a substance composed mainly of fatty acids and flavonoids obtained from grapefruit seeds), pimento extract (defined as a substance composed mainly of eugenol and thymol obtained from pimento fruits) and purple yam colour (defined as a substance composed mainly of cyanidine acylglucosides obtained from yam tuberous roots). A number of enzymes are included in the list, including hemicellulases, pectinases, chitosanase and xylanase. Such additives are generally understood to be acceptable for food use unless otherwise restricted by the Standards for Use of Food Additives.

In contrast, some additives detailed in the List of Designated Food Additives are permitted only in named foods to the maximum limits stated by the Standards for Use of Food Additives (for example, sorbic acid is permitted to max. 1000 mg/kg in syrups, and calcium disodium EDTA to max. 35 mg/kg in canned or bottled non-alcoholic beverages). In other cases, where no restrictions are given, for example with lecithin and glycerol esters of fatty acids, it is understood that such additives are generally permitted unless restricted or prohibited by a standard of composition. As there are few compositional standards in a European sense, with standards mainly relating to hygiene, such additives tend to be generally acceptable. In some cases, named additives (for example permitted colours) are not permitted in a range of basic foods, but are otherwise generally permitted. The List of Designated Food Additives includes compounds permitted for flavouring (of the synthetic type) and as dietary supplements.

Specific standards are established for the majority of additives in the List of Existing Food Additives in a separate publication from the Food Sanitation Law entitled 'Japan's Specifications and Standards for Food Additives' (10).

Labelling

In common with requirements in other countries, the majority of food additives need to be declared on the label. Some must be declared by specific name and class name – for example colorants, bleaching agents and antifungal agents; other additives should be declared by specific name, although a limited number of generic terms may be used instead of the specific names, such as seasoning, flavouring, gum base and bittering agent. Processing aids, nutrients/dietary supplements and additives that are present in the final food as a result of carry-over need not be declared. One interesting characteristic of additives labelling requirements in Japan is that the labelling of natural additives should not be discriminated from that of chemical synthetic additives and, hence, the use of the term 'natural' or any equivalent term implying 'natural' is not permitted for food additives.

Other Far East Countries

There is a variety of formats in food additives legislation in the Far East countries (11); as they are major export markets it is worth examining some of the differences between them.

In Malaysia, detailed food standards established in the Food Regulations 1985 control the use of additives. There are a number of lists of permitted additives, including colours, preservatives, nutritive supplements and food conditioners. This latter category covers emulsifiers, stabilisers, thickeners, solvents, acidity regulators, modified starches, gelling agents, enzymes, anti-caking agents and anti-foaming agents. Additives in Singapore are also controlled by detailed food standards, laid down as part of the Food Regulations 1988; lists of permitted antioxidants, colours, emulsifiers and stabilisers, preservatives, anti-caking agents, anti-foaming agents, sweeteners, flavour enhancers, humectants, sequestrant, nutrient supplements and general-purpose food additives are included.

In contrast, South Korea has established a different system of controlling food additives. Classification is by four different categories according to provisions of use:

1. Additives only permitted in certain listed foods for specific functions, e.g. polyvinyl acetate may be used only as a gum base in chewing gum and as a film agent for use on the skin of fruit and vegetables.

2. Additives permitted in certain foods whose function is not specified in the regulations, e.g. potassium sorbate, sodium saccharin.

3. Additives intended for use for a specified function with no restriction on their use in particular foods, e.g. flavours.

4. Additives with no restrictions on their use generally or for particular functions; generally, their use in standardised foods and non-standardised foods is acceptable (Table 4.III).

TABLE 4.III
Examples of Group 4 additives in South Korea

Acetic acid	Carrageenan
Arabic gum	Cyclodextrin
Ferric phosphate	Lysozyme
Folic acid	Methyl hesperidin
Gelatin	L-Phenylalanine
Hemicellulase	Fungal protease
Lecithin	Riboflavin
D-sorbitol	Vitamin A in oil

The use of additives in China is controlled by the National Standard GB 2760-1996 on Hygienic Standard for Uses of Food Additives. Additives may be used in foods only if specifically permitted by the Standard, which also regulates flavours and processing aids. Certain additives are generally permitted, including monosodium glutamate, β-carotene, gelatin, agar and carrageenan. Aspartame is also generally permitted (except in canned food) and additional specific labelling provisions apply. There is specific legislation laid down in Hong Kong for preservatives, antioxidants, colours and sweeteners, but currently, there are no specific provisions laid down for the use of other classes of additive in foods in Hong Kong. The competent authority makes reference to the Codex General Standard on Food Additives, and those set by other countries, in determining whether a food additive is fit for human consumption. In Taiwan, additives are regulated by the Scope and Application Standards on Food Additives, which give positive lists of various additives including flavourings, solvents and processing aids. In Thailand, the use of food additives, except sweeteners and flavourings, in foodstuffs must be in accordance with the Codex General Standards on Food Additives. The use of sweeteners is controlled by the regulations on individual foods. In Vietnam, the Ministry of Health Decision 3742/2001/QD-BYT controls the use of additives in foodstuffs. Further requirements may be established in the food standards.

MERCOSUR

Argentina, Brazil, Uruguay and Paraguay signed the Treaty for the Organisation of a Southern Common Market (MERCOSUR) in 1991 (12). Venezuela signed a membership agreement with MERCOSUR on 17 June 2006, but before becoming a full member, its entry must be ratified by the Paraguayan and Brazilian Parliaments. The aim of MERCOSUR is to create a free trade area and a customs union, together with legislation to avoid distorting operation of a common market. To facilitate trade, harmonised food legislation is being developed by the common market group; each Member State must comply with the provisions of the Asuncion Treaty and implement harmonised MERCOSUR legislation.

A general list of food additives was first established in 1993, however, this has recently been revoked and replaced by a new list of additives, this list also includes colours and sweeteners. In 1996, MERCOSUR published a list of general GMP permitted additives which is still in force. The general list of permitted additives is not intended to be a list of generally permitted food additives. Until such provisions are established, the individual Members of MERCOSUR will establish their own conditions of use, unless additives were covered by a specific MERCOSUR standard. It is, therefore, the national law that still has to be considered. However, recently, new resolutions have been published that detail lists of additives and their levels permitted in specific food categories – for example, non-alcoholic beverages, chocolate, cereal and cereal-based products, and sugar confectionery. The MERCOSUR Member States vary in the time they take to implement these resolutions into their national legislation; for example, Argentina and Brazil implement all the resolutions on additives promptly, but other Member States may take several years to do so. Ultimately, it is intended that there will be a harmonised list of food additives by individual resolutions on commodities. It is apparent that such a system will take some time to implement fully, and one of the problems in such a system is trying to classify the range of potential food products that may be produced.

The general list classified the additives listed by functional class, listing them using the Codex INS number. Included among the list are compounds such as nitrates, ascorbates, lactates and phosphates, generally regulated in most countries' food law.

Middle East

Key legislation in the context of the Middle East includes that of Saudi Arabia and the Gulf States (13). Saudi regulations exist on colours, preservatives, antioxidants, flavourings and emulsifiers, stabilisers and thickeners, as well as on maximum levels for the use of benzoic acid, benzoates and sulphites in a range of foods. These regulations generally list the permitted additives but do not specify in which foods the additives may be used; individual food standards contain relevant provisions concerning specific foods. In the absence of specific authorisation in a food standard for use of an additive, it is necessary to contact the authorities to obtain approval for its use. This can cause problems for manufacturers of products that are not standardised as individual approvals are required, which can be time-consuming. With respect to sweeteners, a Gulf Standard has now been adopted as the Saudi standard. Import (and use in Saudi Arabia) of alcohol (including foods prepared or preserved with alcohol) and cyclamates is not permitted. It is, therefore, best to use alternative solvents to alcohol-based compounds when exporting to the Middle East.

A similar pattern is seen with Gulf Standards, adopted by the Council of the Gulf Co-operation Council countries. There are standards on permitted additives in edible oils and fats, colouring matter used in foods, permitted preservatives, antioxidants and emulsifiers, stabilisers and thickeners, which list permitted additives but where, again, it is the individual food standards that are applicable in terms of additive acceptability. Gulf standards on salts of sulphurous acids and benzoates used in the preservation of foods and on sweeteners also specify foods in which these additives may be used. Artificial sweeteners are permitted only in foods for particular nutritional use, such as diabetic food, or in energy-reduced foods or foods with no added sugars. It is still recommended that the authorities or a local agent be contacted to confirm the approval of sweeteners.

Additives legislation in Israel is based upon European Parliament and Council Directive 94/36/EC on colours for use in foodstuffs, European Parliament and Council Directive 94/34/EC, as amended, on sweeteners for use in foodstuffs, and European Parliament and Council Directive 95/2/EC, as amended, on food additives other than colours and sweeteners, implemented as The Public Health (Food Additives) Regulations 2001. The annex to the Regulation lists all food additives permitted for use in the various foodstuffs. The majority of provisions on the use of additives and maximum levels laid down in the EC Directives are maintained in the Israeli Regulations, including the principles of *Quantum Satis* and Carry-Over, with some alterations or amendments made to a few specified additives or categories of foodstuffs in which the additives are permitted for use. The provisions of Israeli compositional standards relating to the use of additives are superseded by the additives regulations, except for provisions on the use of additives in 13 foodstuffs typical for Israeli consumption, namely: Sesame halvah, soft white cheese, fermented milk products, grape juice, sesame tehina, pickled black olives, preserved grape juice, specific salads, mixed spices, salty cheeses and matzoth for Passover. The Regulations and annex are available in Hebrew only and no official English translation will be issued. The Regulations are based upon the English version of the EC Directives and translated in Hebrew, therefore translation back into English may result in errors of interpretation. It is recommended that the EC Directives themselves be used for guidance.

The use of additives in raw materials used for the professional, industrial manufacture of foodstuffs are not covered by the Public Health (Food Additives) Regulations; Israeli standards apply.

An additional complication for additives in Israel is that certain additives may not be Kosher and this may give rise to problems – for example, the use of animal-derived additives in a dairy product.

Australia and New Zealand

The use of food additives in both Australia and New Zealand is regulated by the Australia New Zealand Food Standards Code (Code) and enforced in both countries under State and Territory food laws. The use of additives in specific foods is detailed in Standard 1.3.1.

Food Standards Australia New Zealand (FSANZ) is responsible for the development of, or variation to, food standards established in the Code. FSANZ is an independent statutory agency established by the *Food Standards Australia New Zealand Act 1991*. Working within an integrated food regulatory system involving the governments of Australia and the New Zealand Government, it set food standards for the two countries. FSANZ is part of the Australian Government's Health and Ageing portfolio (14).

In New Zealand, the New Zealand Food Safety Authority (NZFSA) is responsible for the implementation of the Australia New Zealand Food Standards Code which took full effect on 20 December 2002. Food sold in New Zealand must comply with the Code.

The standard development process involves an evaluation of the risk to public health of the proposed change to the Code and the impact of the regulatory measures on the food industry and our international trading obligations. Then FSANZ draft a legal standard for public comment. There may be one or more periods of public consultation for each standard.

Finally, the draft standard is considered for approval by the FSANZ Board and, if the Ministerial Council does not request a review of the decision within 60 days, FSANZ gazette publish the standard as law.

Before the use of a new additive in a food can be approved, it is necessary to have answers to the following questions:

*   Is the additive safe to eat at the requested level in a particular food?
*   Are there good technological reasons for the use of the additive?
*   Will consumers be clearly informed about its presence?

Only if satisfaction is reached on these points will a recommended maximum permitted level in a particular food be put forward, based on technological need and provided it is well within safety limits. FSANZ allows additives to be used only if it can be demonstrated that no harmful effects are expected, following evaluation of data obtained by testing the additive. Food additive safety is based on the ADI. A review of individual food standards is currently underway to bring them up to date with the modern food industry. Dietary evaluation has been carried out to ensure that consumption is well within safe limits, even for those who consume large quantities of certain foods. It is considered that the revised standard is more flexible for industry, allowing technological developments while maintaining product safety.

The Australian New Zealand Food Standards Code is unusual in that it contains Standard 1.3.3 on processing aids. In many countries, processing aids are not specifically regulated; under this Standard the specific compounds that may be used for processing aid functions are detailed. The Code defines a processing aid as a substance used in the processing of raw materials, foods or ingredients to fulfil a technological purpose relating to treatment or processing, but which does not perform a technological function in the final food. In contrast to the situation in most countries, the use of processing aids in foods is prohibited unless specific provision is given for that use. The processing aids categories covered by the Standard include generally permitted processing aids; catalysts; bleaching agents, washing and peeling agents; extraction solvents; enzymes; microbial nutrients and microbial nutrient adjuncts; and those processing aids used in packaged water and in water used as an ingredient in other foods.

TABLE 4.IV
Examples of permitted processing aids under the Australia New Zealand Food Standards Code

Processing aid	Maximum residue (mg/kg)
White mineral oil	Generally permitted
Hydrogen peroxide	5
Sodium hypochlorite	1 (available chlorine) as a bleaching agent, washing or peeling agent
Methanol	5 as an extraction solvent
β-cyclodextrin	Good Manufacturing Practice used to extract cholesterol from eggs
Chlorine dioxide in packaged water and water used as a food ingredient	5 (available chlorine)

Enzymes of microbial, animal or plant origin are detailed, together with their authorised sources. For example, Bromelain from pineapple stem (Ananas comosus) and alpha-Galactosidase from Aspergillus niger are listed.

Labelling

Standard 1.2.4 of the Code covers provision on declaration of additives. Declaration of additives in the list of ingredients is by class name for some 28 classes of additive, together with, in brackets, the prescribed name, and appropriate designation or code number of each additive belonging to the class that is or may be present in the food. The code numbers are as listed under EC additives legislation but without the 'E' prefix. As in Europe, it is the function of the additive in the food that determines additive class declaration; if an additive can perform more than one function, then the one most appropriate to the function being carried out in the particular food must be declared.

References

1   CODEX Alimentarius Commission, Procedural Manual, 16th edition, FAO/WHO.

2   CODEX General Standard for Food Additives CODEX STAN 192-1995 (amendments to 2007).

3   Joint FAO/WHO Food Standards Programme, Codex Alimentarius Commission, Report of the 39th Session of the Codex Committee on Food Additives, Rome , Italy, 2007.

4.   Wyles A.M. Food Legislation Surveys No. 2, An Index to Food Additives evaluated by the Joint FAO/WHO Expert Committee on Food Additives. 6th Edn. Leatherhead, LFRA Ltd. 1998.

5.   FDA/CFSAN. Food Additive Petition Expedited Review – Guidance for industry and CFSAN Staff. Web site, http://vm.cfsan.fda.gov/~dms/opa-expe.html

6.   FDA/CFSAN. Web site, http://vm.cfsan.fda.gov

7.   Federal Register, 62, (74), 17/4/97, 18937–64.

8.   GRAS notification status. Web site, http://vm.cfsan.fda.gov/~dms/opa-gras.html

9.   Specifications and Standards for Foods, Food Additives, etc. under the Food Sanitation Law JETRO, January 2003.

10.   Japanese Standards for Food and Additives. 7th Edn. Ministry of Health and Welfare, 2000. [Note: We understand that an 8th Edition was published in September 2007. This is not available outside Japan at the time of going to press.]

11.   Holden J., Outen E., Thompson R., Kardos H. Food Legislation Surveys No. 7, Guide to Food Regulations in the Far East and South Asia. 4th Edn. Leatherhead, LFRA Ltd. 1998, updated to September 1999.

12.   Horne I., Navarro S., Thompson R. Food Legislation Surveys No. 9, Guide to Food Regulations in Latin American Countries. Leatherhead, LFRA Ltd. 1998.

13.   Wood R., Morgan C. and Legislation Unit. Food Legislation Surveys No. 4, The Middle East. 4th Edn. Leatherhead, LFRA Ltd. 1992, updated to July 1999.

14   Food Standards Australia New Zealand (FSANZ), http://www.foodstandards.gov.au

5. ADDITIVES

E100	Curcumin
	CI natural yellow 3
	Turmeric yellow
	Diferoyl methane

Colour Index No: 75300

Sources
Curcumin is the principal pigment of turmeric, a spice that is obtained from the rhizomes of *Curcuma longa*. It is obtained by solvent extraction from the plant to produce an oleoresin, which is then purified by crystallisation.

Function in Food
Curcumin is a natural colour that provides a bright lemon-yellow colour when used in foods. Although oil-soluble, it is available in water-dispersible forms.

The pure pigment has a high tinctorial strength with an absorption maximum in the region of 426 nm when measured in acetone.

Benefits
Curcumin is a component of turmeric, a spice that has been consumed for many thousands of years. It has an ADI of 0-3 mg/kg body weight allocated by JECFA and is an approved colour for use in foodstuffs according to Directive 94/36/EC. It is listed in Annex V Part 2 as a colour that may be used in a wide range of foods subject to specific quantitative limits.

Its stability to heat is excellent and it may generally be used in products throughout the acid pH range.

Limitations
Curcumin has poor stability to light and is sensitive to sulphur dioxide at levels in excess of 100 parts per million (ppm).

Typical Products
Curcumin provides a lemon-yellow hue and is widely used to colour smoked white fish, ice creams, sorbets, dairy products and some types of sugar confectionery. Typical dose applications calculated on the basis of the pure pigment are between 10 and 100 parts per million.

E101	Riboflavin
	(i) Lactoflavin
	Vitamin B$_2$
	Riboflavin-5'-phosphate
	(ii) Riboflavin-5'-phosphate, sodium

Sources

Riboflavin is found in green vegetables, milk, eggs and yeast. It is produced commercially as yellow crystals by chemical synthesis. Riboflavin 5'-phosphate sodium is a more water-soluble form.

Function in Food

Riboflavins are used to provide a bright lemon yellow, but generally the colour is incidental to the use as a vitamin.

Benefits

Both forms are nutritionally important as vitamin B$_2$ and the coloration properties are often of secondary importance. They have good stability to heat and moderate stability to acid pH. JECFA has allocated ADIs of 0.5 mg/kg body weight. They are listed in the EC Colours Directive 94/36/EC in Annex V Part 1 and hence can be used to *quantum satis* in a wide variety of foodstuffs.

Limitations

Stability to light when in solution is poor and so applications in aqueous solution should be protected from light. Riboflavin is only slightly soluble in water and riboflavin-5'-phosphate should be used where solubility is an important factor. Both can impart a bitter taste.

Typical Products

Because both forms are affected by UV light, their use as a colour is restricted to salad dressings, confectionery, tablet coatings and powdered drinks.

E102	Tartrazine
	CI food yellow 4
	FD&C yellow no. 5

Colour Index No: 19140

Sources

A water-soluble synthetic dye commercially available as the sodium salt of the dye. Tartrazine is also available as the aluminium lake, which is water-insoluble. Tartrazine is classed as a monoazo dye.

Function in Food

Tartrazine is a synthetic colour that is used to provide a light, typically lemon-yellow colour to foods, particularly those of lemon and lime flavours. The colour is lemon yellow but less green than quinoline yellow. It is often used in combination with other colours.

The pure pigment has a high tinctorial strength with an absorption maximum in water of 426 nm.

Benefits

It has good stability to heat and light and has been allocated an ADI of 7.5 mg/kg body weight by JECFA. It is approved to colour foodstuffs according to Directive 94/36/EC and is listed in Annex V Part 2 as a colour that may be used in a wide range of foods subject to specific quantitative limits. It is an approved colour in the USA where it is certified by the FDA as FD&C yellow no. 5.

Limitations

In general use, tartrazine does show some instability to ascorbic acid and fades in the presence of sulphur dioxide at concentrations higher than 50 ppm.

Typical Products

Soft drinks, canned foods, edible ices, desserts, confectionery, pickles, sauces and seasonings.

E104	Quinoline yellow
	CI food yellow 13

Colour Index No: 47005

Sources

A water-soluble synthetic dye commercially available as the sodium salt of the dye. Quinoline yellow is also available as the aluminium lake, which is water-insoluble. Quinoline yellow is classed as a quinophthalone dye.

The dye consists essentially of a mixture of principally the disulphonates and to a lesser proportion the monosulphonates of the dye.

Function in Food

Quinoline yellow is a synthetic colour that is used to provide a greenish-yellow, simulating the shade of pineapple and lemon. It is often used in combination with other colours.

The pure pigment has a high tinctorial strength with an absorption maximum in water of 411 nm.

Benefits

It is approved to colour foodstuffs according to Directive 94/36/EC, and has been allocated an ADI of 10 mg/kg body weight by JECFA. In the Directive it is listed in Annex V Part 2 as a colour that may be used in a wide range of foods subject to specific quantitative limits.

The dye has good stability to heat and light and is generally stable in the presence of fruit acids and sulphur dioxide.

Limitations

It is somewhat unstable in alkaline conditions and will precipitate in the presence of benzoic acid. Quinoline yellow is not permitted in the USA and should not be confused with D&C yellow no. 10. Whilst both have the same colour index number, it does not contain the same proportion of sulphonated components. D&C yellow no. 10 is predominantly the monosulphonate of the dye, and is furthermore not authorised for use in food in the USA.

Typical Products

Soft drinks, jams and preserves, edible ices, desserts, confectionery, pickles, sauces and seasonings.

E110	Sunset yellow FCF
	CI food yellow 3
	Orange yellow S
	FD&C yellow no. 6
	Colour Index No: 15985

Sources

A water-soluble synthetic dye commercially available as the sodium salt of the dye. Sunset yellow is also available as the aluminium lake, which is water-insoluble. Sunset yellow is classed as a monoazo dye.

Function in Food

Sunset yellow is a synthetic colour that is used to provide an orange shade characteristic of orange peel. It is often used in combination with other colours. The pure pigment has a high tinctorial strength with an absorption maximum in water of 485 nm.

Benefits

It has good stability to heat and light and is approved to colour foodstuffs according to Directive 94/36/EC. It is listed in Annex V Part 2 as a colour that may be used in a wide range of foods subject to specific quantitative limits. It is an almost globally approved colour, including in the USA, and is certified by the FDA as FD&C yellow no. 6. It has been allocated an ADI of 2.5 mg/kg body weight by JECFA.

Limitations

Although it is generally stable in solution between pH levels 3-8, it is only moderately stable in the presence of benzoic acid and sulphur dioxide, fading in the latter.

This colour is commonly used in soft drinks, but does have a tendency to precipitate in the presence of calcium ions from the gums used, as the less soluble calcium salt. Additions of permitted levels of subsidiary dyes reduce this problem.

Typical Products

Soft drinks, jams and preserves, edible ices, desserts and confectionery, pickles, sauces and seasonings.

E120	Cochineal
	Carminic acid
	Carmine

Colour Index No: 75470

Sources

Carminic acid is the red pigment obtained by aqueous alkaline extraction from the dried bodies of the coccid insect (*Coccus cacti* L.). The word cochineal is used to describe both the dried insects and the colour extracted from them.

Carmine pigment is the aluminium lake of carminic acid.

Function in Food

Carmine has a long history of use as a food colour that provides a bright strawberry red shade to a wide variety of products. It is generally used in products in which the pH is above 3.5 and is available in both water-insoluble and water-soluble forms.

Carminic acid is water-soluble and is particularly appropriate for providing clear orange hues in acid-based preparations such as soft drinks.

The absorption maximum for carmine is in the region of 518 nm when measured in aqueous ammonia solution. Carminic acid has an absorption maximum in the region of 494 nm in dilute hydrochloric acid.

Benefits

Both carminic acid and carmine are chemically very stable, with excellent resistance to oxygen, light, sulphur dioxide, heat and water activity. They are both allocated an ADI by JECFA and SCF of 5 mg/kg body weight per day and are approved colours for use in foodstuffs according to Directive 94/36/EC. They are listed in Annex V part 2 as colours that may be used in a wide range of foods subject to specific quantitative limits.

Limitations

Carmine precipitates in low-pH conditions and should not be used in foods for which the pH is below 3.5. The shade of carminic acid is dependent upon pH, and it is generally suitable only for the coloration of acidic products.

Neither carmine nor carminic acid may be used in products claiming their suitability for vegetarian diets.

Typical Products

Cochineal and its derivatives are used to colour meat products, beverages, table jellies, sugar confectionery, yoghurts and desserts. Typical dose applications calculated on the basis of the pure pigment are between 5 and 50 ppm.

E122	**Carmoisine**
	Azorubine
	CI food red 3
	Colour Index No: 14720

Sources

A water-soluble synthetic dye commercially available as the sodium salt of the dye. Carmoisine is also available as the aluminium lake, which is water-insoluble. Carmoisine is classed as a monoazo dye.

Function in Food

Carmoisine is a synthetic dye used to provide a bluish red colour appropriate for raspberry- or blackcurrant-flavoured foods. It is less blue than amaranth and is used in combination with other colours to provide the desired hue.

The pure pigment has a high tinctorial strength with an absorption maximum in water of 516 nm.

Benefits

Carmoisine is particularly useful for its stability to light. In Directive 94/36/EC it is listed in Annex V Part 2 as a colour that may be used in a wide range of foods subject to specific quantitative limits. It has been allocated an ADI of 0-4 mg/kg body weight by JECFA.

Limitations

Carmoisine has no technical limitations to its use.

Typical Products

Soft drinks, edible ices, desserts and confectionery.

E123	Amaranth
	CI food red 9

Colour Index No: 16185

Sources

A water-soluble synthetic dye commercially available as the sodium salt of the dye. Amaranth is also available as the aluminium lake, which is water-insoluble. Amaranth is classed as a monoazo dye.

Function in Food

Amaranth is a synthetic dye used to provide a deep bluish red shade typical of red berry foods. The pure pigment has a high tinctorial strength with an absorption maximum in water of 520 nm.

Limitations

Under Directive 94/36/EC amaranth is permitted only in certain apéritif wines, spirit drinks and fish roe.

Amaranth is no longer authorised for use in the USA.

E124	Ponceau 4R
	Cochineal red A
	CI food red 7
	New coccine

Colour Index No: 16255

Sources

A water-soluble synthetic dye commercially available as the sodium salt of the dye. Ponceau 4R is also available as the aluminium lake, which is water-insoluble. Ponceau 4R is classed as a monoazo dye.

Function in Food

Ponceau 4R is a synthetic dye used to provide a bright red shade typical of strawberry-, cherry- or redcurrant-flavoured foods. It is used in combination with other colours to provide the desired hue. The lake colour is used in the coloration of cheese rind. The pure pigment has a high tinctorial strength with an absorption maximum in water of 505 nm.

Benefits

It is approved to colour foodstuffs according to Directive 94/36/EC, and has been allocated an ADI of 4 mg/kg body weight by JECFA. In Directive 94/36/EC it is listed in Annex V Part 2 as a colour that may be used in a wide range of foods subject to specific quantitative limits.

A particularly bright red colour which has good light and heat stability.

Limitations

Restricted inclusion rates when used in non-alcoholic flavoured drinks, edible ices, desserts fine bakery wares and confectionery.

Typical Products

Soft drinks, edible ices, confectionery, desserts and cheese.

E127	**Erythrosine**
	CI food red 14
	FD&C red no. 3
	Colour Index No: 45430

Sources

A water-soluble synthetic dye commercially available as the sodium salt of the dye. Erythrosine is also available as the aluminium lake, which is water-insoluble. Erythrosine is classed as a xanthene dye. It is manufactured by the treatment of boiling alcoholic solution of fluorescein with iodine and sodium iodide.

Function in Food

Erythrosine is a synthetic dye used to impart a light red colour to foods. The pure pigment has a high tinctorial strength with an absorption maximum in water of 526 nm.

Benefits

A unique bright pink shade with good stability to heat, and highly stable in the presence of sulphur dioxide. The dye is particularly suitable for colouring maraschino and canned cherries, where no bleeding is essential. The dye is precipitated in the fruit by the presence of the fruit acids. Erythrosine has been allocated an ADI of 0.1 mg/kg body weight by JECFA.

Limitations

In acidic sugar confectionery, erythrosine undergoes colour change to an orange shade. Stability to light is poor and it is precipitated in acidic conditions to its free acid. The erythrosine molecule contains iodine, which has been linked to thyrotoxicosis, as a result of which the use of the dye has been restricted in its applications. According to Directive 94/36/EC, erythrosine is permitted only for colouring cherries.

E129	**Allura red AC**
	CI food red 17
	FD&C red no. 40
	Colour Index No: 16035

Sources

A water-soluble synthetic dye commercially available as the sodium salt of the dye. Allura red is also available as the aluminium lake, which is water-insoluble. Allura red is classed as a monoazo dye.

Function in Food

Allura red provides an orange-red shade in solution, somewhat weaker in strength than the other reds. In the USA, it produces the characteristic shade of blueberries. The pure pigment has a medium tinctorial strength with an absorption maximum in water of 504 nm.

Benefits

Allura red is a general-purpose colour with reasonable stability in a variety of foods and tolerance to processing and storage. It is permitted in Annex V Part 2 of Directive 94/36/EC for use in a wide range of foods and has been allocated an ADI of 7 mg/kg body weight by JECFA.

Limitations

It is not very stable in the presence of oxidising and reducing agents and tends to go bluer in alkaline conditions. It does not provide a good dye for blending purposes; mixtures containing allura red tend to be dull. As a somewhat weak shade, it needs to be used at concentrations of around 100 mg/kg to produce an adequate coloration.

Typical Products

Soft drinks, confectionery, edible ices and desserts.

E131	**Patent blue V**
	CI food blue 5

Colour Index No: 42051

Sources

A water-soluble synthetic dye commercially available as either the sodium or calcium salt of the dye. Patent blue V is also available as the aluminium lake, which is water-insoluble. Patent blue V is classed as a triarylmethane dye.

Function in Food

Patent blue V is a bright blue colour but is mainly used in combination with tartrazine or quinoline yellow to provide a green colour. The only significant use of the colour alone is in carcass staining.

Benefits

Highly stable colour to heat and light; less stable to food processing ingredients, such as fruit acids and benzoic acid.

Limitations

Although available in commercial quantities, this dye has a limited number of manufacturers. It should not be confused with the dye blue VRS, which has a very similar molecular structure, but is not a permitted food dye.

It is permitted for use under Annex V Part 2 of Directive 94/36/EC. Patent blue V has not been allocated an ADI by JECFA, although the SCF has assigned a figure of 15 mg/kg body weight.

Moderate solubility in water is a limitation but, at a maximum of 4-6% at room temperature, makes it suitable for use at levels of 50–200 mg/kg.

Typical Products

Baked goods and confectionery.

E132	Indigo carmine
	Indigotine
	CI food blue 1
	FD&C blue no. 2

Colour Index No: 73015

Sources

A water-soluble synthetic dye commercially available as the sodium salt of the dye. Indigo carmine is also available as the aluminium lake, which is water-insoluble. Indigo carmine is classed as a indigoid dye.

Function in Food

Indigo carmine is used to provide a dark bluish-red colour to foodstuffs. It is often used in combination with other colours.

The pure pigment has a high tinctorial strength with an absorption maximum in water of 610 nm.

Benefits

Indigo carmine has wide acceptability and has been allocated an ADI by JECFA of 5 mg/kg body weight. It is an approved colour in the USA when certified by the FDA as FD&C blue no. 2. It is included in Annex V part 2 of Directive 94/36/EC, where it is permitted in a wide range of foodstuffs with individual maxima in each case.

Limitations

The instability of this dye is its most limiting factor. It is unstable to most conditions – heat, light and common food ingredients. It fades in the presence of sulphur dioxide and sugar and syrup solutions. Fading increases with pH across the range pH 5 to pH 8 and fades completely at pH 9.

Its limited solubility in water at 1–2% also restricts its use. This dye can be used at concentrations of 50–200 ppm in water.

Typical Products

Although poorly stable, its wide permissibility makes it a commonly used colour. It is often used in combination with other dyes depending on its use limitations. Typical products include confectionery, baked goods and edible ices.

E133	**Brilliant blue FCF**
	CI food blue 2
	FD&C blue no. 1

Colour Index No: 42090

Sources

A water-soluble synthetic dye commercially available as the sodium salt of the dye. Brilliant blue is also available as the aluminium lake, which is water-insoluble. Brilliant blue is classed as a triarylmethane dye.

Function in Food

Brilliant blue provides a greenish-blue colour, which is particularly used for blending with tartrazine or quinoline yellow to give greens and with other colours to give browns and blacks. The pure pigment has a high tinctorial strength with an absorption maximum in water of 630 nm.

Benefits

Brilliant blue is a very stable colour and is widely used. It has been allocated an ADI of 12.5 mg/kg body weight by JECFA and is permitted in Directive 94/36/EC where it is included in Annex V Part 2. It is permitted in the USA as FD&C blue no. 1.

Limitations

Brilliant blue tends to fade at pH 8 and above.

Typical Products

Soft drinks, canned and baked goods, confectionery, desserts and edible ices.

E140	(i)	**Chlorophyll** **CI natural green 3** **Magnesium chlorophyll** **Magnesium phaeophytin**
		Colour Index No: 75470
	(ii)	**Chlorophyllin** **CI natural green 5** **Sodium chlorophyllin** **Potassium chlorophyllin**
		Colour Index No: 75815

Sources

The oil-soluble chlorophylls are extracted from edible plant material including grass, lucerne and nettle. Alkaline saponification produces the chlorophyllins, which are soluble in water.

Function in Food

Chlorophylls and chlorophyllins are naturally derived colours that provide green hues to food products. Commercially, products are available for use in both oil- and water-based systems.

The absorption maximum of chlorophyll is in the region of 409 nm, measured in chloroform, and that of the chlorophyllins around 405 nm, measured in aqueous pH 9 buffer solution.

Benefits

Chlorophylls are natural pigments present in all green leafy vegetation, and they have always been a component of man's diet. They do not have a specified ADI according to JECFA, and are approved colours for use in foodstuffs according to Directive 94/36/EC. They are listed in Annex V part 1 as colours that may be used at *quantum satis*.

Limitations

Providing dull olive-green hues, the chloropylls are less stable to light and acidic conditions than their coppered counterparts (E141).

Typical Products

Chlorophylls have only limited application in foodstuffs. They may be used in sugar confectionery, yoghurts and ice cream. Typical dose applications calculated on the basis of the pure pigment are between 20 and 200 ppm.

E141	(i)	**Copper complexes of chlorophylls** **CI natural green 3** **Copper chlorophyll** **Copper phaeophytin**
	(ii)	**Copper complexes of chlorophyllins** **CI natural green 5** **Sodium copper chlorophyllin** **Potassium copper chlorophyllin**
		Colour Index No: 75815

Sources

Copper chlorophylls are obtained by addition of a copper salt to extracted edible plant material including grass, lucerne and nettle. Alkaline saponification prior to reaction with a copper salt produces the copper chlorophyllins, which are soluble in water.

Function in Food

Copper complexes of chlorophylls and chlorophyllins are chemically modified natural extracts that are used as colours providing blue-green hues to food products. Commercially, products are available for use in both lipid and aqueous media.

The absorption maximum of both products is in the region of 405–425 nm when measured, respectively, in chloroform or aqueous pH 7.5 buffer solution.

Benefits

Copper chlorophylls provide brighter and more stable colours than their uncoppered counterparts. The ADI allocated by both JECFA and SCF is 15 mg/kg body weight, and they are approved colours for use in foodstuffs according to Directive 94/36/EC. They are listed in Annex V part 1 as colours that may be used at *quantum satis*.

Limitations

As chemically modified extracts, the copper chlorophylls should not be described as natural colorants. They are approved for use in the EU, however, in the USA, only sodium copper chlorophyllin may be used to colour citrus-based dry beverage mixes in an amount not exceeding 0.2% in the dry mix.

Typical Products

Copper chlorophylls are widely used in sugar confectionery, yoghurts, ice cream, sauces, pickles and jams. Typical dose applications calculated on the basis of the pure pigment are between 20 and 200 ppm.

E142	**Green S**
	CI food green 4
	Brilliant green BS
	Colour Index No: 44090

Sources

A water-soluble synthetic dye commercially available as the sodium salt of the dye. Green S is also available as the aluminium lake, which is water-insoluble. Green S is classed as a triarylmethane dye.

Function in Food

A greenish-blue colour in solution, characteristically more blue than its name suggests. It is most commonly used to produce green shades with tartrazine and quinoline yellow. Green S is traditionally the colour used for canned peas and also provides a good shading colour in combination with other colours to produce browns and black shades. The pure pigment has a high tinctorial strength with an absorption maximum in water of 632 nm.

Benefits

Green S is permitted in Directive 94/36/EC, where it is included in Annex V Part 2 and permitted in a wide range of food products. It has been allocated an ADI of 5 mg/kg body weight by the SCF.

Limitations

This dye has very few manufacturers in the world, being complex in structure and a difficult dye to synthesise.

Typical Products

Canned peas and soft drinks.

E150a	Class I	**Plain caramel**
E150b	Class II	**Caustic sulphite caramel**
E150c	Class III	**Ammonia caramel**
E150d	Class IV	**Sulphite ammonia caramel**

Sources

The caramelisation of sugar is well known in cooking. The production of plain caramel is an extension of this process, heating sugars (glucose, fructose or sucrose) to elevated temperature under pressure. The other classes of caramel are produced using alkalies, ammonia, ammonium salts or sulphites. The caramels have a range of shades of brown. Caramels are available as both liquids and powders.

Function in Food

Caramels are used to provide brown colours from red through brown to almost black. The taste is bitter, characteristic of burnt sugar. They are typically used for colouring both alcoholic and non-alcoholic drinks.

Benefits

The different production processes result in caramels with different electric charges. Beers and stouts require a caramel with a positive charge to avoid reaction with positively charged compounds derived from malt, while soft drinks use a negatively charged colour, which is more stable in the acidic environment of the drink. Spirits are often coloured with plain caramels. The major use of caramel is in cola drinks. Caramels are totally miscible with water and all four classes are permitted in the EU under Annex V part 1 of Directive 94/36/EC.

Limitations

There are no technical limitations on the use of caramels, although it is important to select the type that is most appropriate for the intended use.

Typical Products

Beer, soft drinks, gravies and sauces, meat products.

E151	Black PN
	Brilliant black BN
	CI food black 1

Colour Index No: 28440

Sources

A water-soluble synthetic dye commercially available as the sodium salt of the dye. Black PN is also available as the aluminium lake, which is water-insoluble. Black PN is classed as a bisazo dye.

Function in Food

Black PN is a blue-black colour on its own but is mainly used for blending to provide violet to purple shades. The pure pigment has a high tinctorial strength with an absorption maximum in water of 570 nm.

Benefits

Black PN is permitted for colouring foodstuffs according to the Directive 94/36/EC, and has been allocated an ADI of 0-1 mg/kg body weight by JECFA. It is listed in Annex V Part 2 as a colour that may be used in a wide range of foods subject to specific quantitative limits.

Limitations

Although moderately stable to light, this dye has poor heat stability and, whilst stable in alkaline conditions, it is not very stable in the presence of fruit acids and sulphur dioxide.

Typical Products

Fish roe products.

E153	Vegetable carbon
	Vegetable black

Colour Index No: 77266

Sources

Vegetable carbon is manufactured by the heating of vegetable material to a high temperature in the absence of air. While wood, cellulose residue, or coconut shell can be used, the main source is peat, as the product from this source tends to have the lowest ash content.

Function in Food

Vegetable carbon is a black powder, insoluble in water or organic solvents, which is used to darken the colour of solid foodstuffs.

Benefits

Vegetable carbon is inert, odourless and tasteless. It is permitted in the EU in Directive 94/36/EC under Annex V Part 1 for use in a wide range of foods to *quantum satis*. It has not been allocated an ADI by either JECFA or the SCF.

Limitations

There are only limited manufacturing sources for this product, which is difficult to handle and use as a powder. It is commonly prepared as a paste or dispersion before incorporation into a product. Although it is widely permitted, it is rarely used since low levels of use tend to produce grey shades, while the level necessary to produce black is often above rates that would be considered Good Manufacturing Practice.

Typical Products

Confectionery, particularly liquorice.

E154   Brown FK
**   CI food brown 1**

Sources

A multi-component water-soluble synthetic dye manufactured as the sodium salts of the dyes present.

Commercial manufacture of Brown FK to meet EC specifications is difficult and producers can only make this dye to non-EC specifications.

Function in Food

Historically, this dye found use in the coloration of smoked herrings to produce kippers. It produces a reddish-brown solution in water.

Benefits

It has strong protein affinity and is stable in brine solutions.

Limitations

According to Directive 94/36/EC, Brown FK was permitted only for colouring kippers. Since it is no longer available to EC specifications, it is no longer used.

Typical Products

Coloration of kippers only.

E155	**Brown HT**
	CI food brown 3
	Chocolate brown HT
	Colour Index No: 20285

Sources

Brown HT is a bisazo water-soluble synthetic dye. It is manufactured as the sodium salt and has also been prepared as the water-insoluble aluminium lake.

Function in Food

A reddish–brown colour in solution, this dye has found most use in the baking industry – hence its suffix, HT, which means "high temperature". The pure pigment has a high tinctorial strength with an absorption maximum in water of 460 nm.

Benefits

Brown HT is soluble in both water and propylene glycol. It is stable to heat and light and to the action of both alkalis and fruit acids. In the EU, it is included in Directive 94/36/EC, where it is permitted in a wide range of foods. It has been allocated an ADI of 1.5 mg/kg body weight by JECFA.

Limitations

Brown HT has no technical limitations to its use.

Typical Products

Baked goods and confectionery.

E160a(i)	**Mixed carotenes** **CI food orange 5**
	Colour Index No: 75130
E160a(ii)	**β-carotene** **CI food orange 5**
	Colour Index No: 40800

Sources

Mixed carotenes are a mixture of natural products obtained by solvent extraction of edible plants and vegetables or from natural strains of the algae Dunaliella salina; β-carotene is the major constituent of this mixture.

The colour β-carotene is a nature-identical pigment that is produced by chemical synthesis or by fermentation using the fungus *Blakeslea trispora*. Although both mixed carotenes and β-carotene are oil-soluble colours, water-dispersible preparations are commercially available.

Function in Food

Mixed carotenes and β-carotene are used to provide yellow and orange shades when used to colour foods.

The pure pigment has a high tinctorial strength, with an absorption maximum in the range 440-457nm when measured in cyclohexane.

Benefits

β-Carotene is a widely distributed carotenoid with a long history of consumption by man. It is a precursor of vitamin A, which thus enables the body to utilise it as pro-vitamin A. It has an ADI of 5 mg/kg body weight allocated by JECFA. Carotenes are approved colours for use in foodstuffs according to Directive 94/36/EC. Both mixed carotenes and β-carotene are listed in Annex V Part 1 as colours that may be used at *quantum satis*.

Stability to heat and pH change is generally good.

Limitations

Carotenes are sensitive to oxidation, especially when exposed to light. Foods and beverages coloured with carotenes frequently benefit from the protective addition of vitamin C and vitamin E derived antioxidants.

Typical Products

Mixed carotenes and β-carotene are used to colour a wide range of foods, including beverages, yellow fats, dairy products and flour confectionery. Typical dose applications calculated on the basis of the pure pigment are between 5 and 50 ppm.

E160b	Annatto
	Bixin
	Norbixin
	CI natural orange 4
	Colour Index No: 75120

Sources

The seeds of the tropical bush *Bixa orellana* L. have long been used as a spice in Central and South America. The bush and the pigment extracted from them are both known as annatto. Bixin is the principal carotenoid pigment extracted from the seeds. Alkaline hydrolysis of bixin converts it from an oil-soluble colour to the water-soluble pigment norbixin.

Function in Food

Annatto is a naturally derived colour that is used to provide orange shades in both lipid and aqueous food phases.

The pure pigment has a high tinctorial strength with characteristic absorption maxima of 470–502 nm for bixin and 452–482 nm in respect of norbixin.

Benefits

Annatto is an approved colour for use in foodstuffs according to Directive 94/36/EC. It is listed in Annexes III and IV as a colour that may be used in a limited range of foods subject to specific quantitative limits.

Its stability to heat is excellent and it may generally be used in products throughout the acid pH range.

Limitations

As a carotenoid, annatto is sensitive to oxidation, especially when exposed to light. The main limitation to the use of this colour in the European market is legislation, restricting its use to a limited range of 15 specific food categories.

Typical Products

Annatto is used to colour yellow fats, cheese, smoked fish, snacks and desserts. Typical dose applications calculated on the basis of the pure pigment are between 10 and 50 ppm.

E160c	**Paprika extract**
	Paprika oleoresin
	Capsanthin
	Capsorubin

Sources

Paprika colour is obtained from sweet red peppers, *Capsicum annuum*, using a solvent extraction process to prepare an oleoresin. Paprika is well recognised as a spice and it is a popular ingredient of many recipe dishes. Although the pigments are oil-soluble, water-dispersible preparations are available commercially.

Function in Food

Paprika extract contains the oil-soluble carotenoid pigments capsorubin and capsanthin. It provides a deep orange hue and imparts a mild spice flavour when used in food products.

The pure pigment has a high tinctorial strength with an absorption maximum in the region of 462 nm when measured in acetone.

Benefits

Paprika colour derives from a spice that has a long history of consumption by man. It does not have a specified ADI according to JECFA and is an approved colour for use in foodstuffs according to Directive 94/36/EC. It is listed in Annex V Part 1 as a colour that may be used at *quantum satis*.

Its stability to heat and pH change is generally good and the mild spice note can be beneficial when it is used in savoury products.

Limitations

Paprika pigments are carotenoids and are sensitive to oxidation, especially when exposed to light. High dose levels may contribute an unacceptable flavour, especially when used in mild-flavoured sweet preparations.

Typical Products

Paprika is used to colour soups, pickles, meat products, sauces, breadcrumbs and snack seasonings. Typical dose applications calculated on the basis of the pure pigment are between 10 and 50 ppm.

E160d   Lycopene
CI natural yellow 27

Colour Index No: 75125

Sources

Lycopene is a carotenoid obtained by solvent extraction of red tomatoes (*Lycopersicon esculentum* L.). It is an oil-soluble pigment, the commercial preparations of which consist of a mixture of carotenoids, with lycopene being the principal constituent.

Function in Food

Commercial preparations of lycopene provide orange and red colours to foods.

The pure pigment has a high tinctorial strength with an absorption maximum in the region of 472 nm when measured in hexane.

Benefits

Lycopene is a natural colour derived by physical means from red tomatoes. Accordingly, it has a long history of consumption by man, and normal dietary intake considerably exceeds that used for the purpose of coloration. It does not have a specified ADI according to JECFA, and is an approved colour for use in foodstuffs according to Directive 94/36/EC. It is listed in Annex V Part 2 as a colour that may be used in a wide range of foods subject to specific quantitative limits.

It is unaffected by pH and exhibits good stability to heat.

Limitations

Lycopene is an oil-soluble carotenoid pigment. It is therefore sensitive to oxidation, especially when exposed to light.

Typical Products

Lycopene may be used to colour soups, sauces and other savoury products. Typical dose applications calculated on the basis of the pure pigment are between 5 and 50 ppm.

E160e   β-apo-8'-carotenal (C30)
**   CI food orange 6**

Colour Index No: 40820

Sources

Although widely distributed in nature, commercial quantities of apocarotenal are chemically synthesised and as such are nature-identical. The pure crystals, like β-carotene, are prone to oxidation, and it is generally available when formulated into oil suspensions or water-dispersible powders.

Function in Food

Imparts a yellow-orange to orange-red colour to foodstuffs. It has an absorption maximum at about 462 nm when measured in cyclohexane.

Benefits

Like β-carotene, apocarotenal has vitamin A activity and has been allocated an ADI of 5 mg/kg body weight by both the SCF and JECFA. It is listed in Annex V Part 2 of the EC Colours Directive 94/36/EC and hence its use is limited to a specified list of foodstuffs where maximum inclusion levels apply. Heat stability is good and the colour is unaffected by changes in pH.

Limitations

Like β-carotene, apocarotenal is prone to photo-oxidation, and products should be protected from UV light or by the presence of an antioxidant such as ascorbic acid. Like other carotenoids, it is not soluble in water, although water-dispersible forms are available.

Typical Products

Apocarotenal can be used, alone or in conjunction with β-carotene, to colour a range of foodstuffs, including soft drinks, confectionery, coatings, ices, soups, desserts and sauces. Its high colour intensity means that relatively low inclusion rates (5-15 ppm) of pure pigment are required.

E160f	Ethyl ester of β-apo-8'-carotenoic acid (C30)
	CI food orange 7
	β-apo-8' carotenic ester

Colour Index No: 40825

Sources

As with other pure carotenoids, this colour is available commercially as a chemically synthesised red to red-violet crystalline powder formulated as a suspension in oil, the pure crystals being prone to oxidation.

Function in Food

Although able to impart a yellow-orange colour, similar to that of β-carotene, to foodstuffs, it is predominantly used as a marker for "intervention" butter and cream. It has an absorption maximum in cyclohexane of 449 nm.

Benefits

The ethyl ester of β-apo-8'-carotenoic acid has been allocated an ADI of 5 mg/kg body weight by both JECFA and the SCF. It is in Annex V Part 2 of the EC Colours Directive 94/36/EC and can be used in specified applications up to specified maximum inclusions. This colour has vitamin A activity (25% that of β-carotene), has good heat stability and is unaffected by changes in pH.

Limitations

Like other carotenoids, the ethyl ester is insoluble in water and, in this case, water-dispersible forms are not currently available.

Typical Products

Although permitted in a number of food applications, this colour is rarely used in the food industry, except as an intervention marker.

E161b	Lutein
	Mixed carotenoids
	Xanthophylls
	Tagetes

Sources

Xanthophylls are a class of carotenoids obtained by solvent extraction from edible fruits and plants, including grass, lucerne (alfalfa) and marigolds (*Tagetes erecta*). They provide an oil-soluble pigment, the commercial preparations of which consist of a mixture of carotenoids, with lutein being the principal constituent.

Function in Food

Oil-soluble and water-dispersible preparations are available commercially to provide a yellow colour to foods in either lipid or aqueous phases.

The pure pigment has a high tinctorial strength with an absorption maximum in the region of 445 nm when measured in chloroform/ethanol or hexane/ethanol/acetone solvent systems.

Benefits

Xanthophylls are a natural colour derived by physical means from edible plants and fruits. Lutein from *T.erecta* has been allocated an ADI of 0-2 mg/kg body weight by JECFA and is an approved colour for use in foodstuffs according to Directive 94/36/EC. They are listed in Annex V Part 2 as a colour that may be used in a wide range of foods subject to specific quantitative limits.

They are unaffected by pH and exhibit good stability to heat.

Limitations

Xanthophyll pigments are carotenoids and are sensitive to oxidation, especially when exposed to light.

Typical Products

Lutein is used commercially in cloudy citrus beverages, sugar confectionery, marzipan and mayonnaise. Typical dose applications calculated on the basis of the pure pigment are between 5 and 50 ppm.

E161g	Canthaxanthin
	CI food orange 8

Colour Index No: 40850

Sources

Found in nature in salmon and trout flesh, crustacea, some fungi and in flamingo feathers, commercial canthaxanthin is chemically synthesised and is hence nature-identical. The deep violet crystals are prone to oxidation and are insoluble in water. An oil suspension and water-dispersible dry forms are available.

Function in Food

Although used widely for the coloration of a range of foodstuffs in Asia and USA, in Europe canthaxanthin can be used only for the coloration of Strasbourg sausages. It provides an orange to violet-red colour.

Benefits

Canthaxanthin exhibits good heat stability and is unaffected by changes in pH.

Limitations

Because of misuse as an artificial tanning aid, in Europe some concerns have been voiced regarding the safety of canthaxanthin. It has been listed in Annex IV of EC Directive 94/36/EC therefore, where it is permitted in "Saucisses de Strasbourg" only, with a maximum permitted level of 15 ppm of pure pigment. For the same reasons, JECFA has allocated a temporary ADI of 0.03 mg/kg body weight.

Typical Products

Canthaxanthin is used for the coloration of "Saucisses de Strasbourg".

E162	Beetroot red
	Beet red
	Betanin

Sources

Beetroot red is obtained from the roots of natural strains of red beets (*Beta vulgaris*) by pressing the crushed beet to express the juice, or alternatively by aqueous extraction of shredded beetroots.

The main colouring principle consists of betacyanins, of which betanin is the major component.

Function in Food

Beetroot red is used to impart a pink colour to foods. Commercial preparations are relatively low in respect of pigment content, although this is partially balanced by the high tinctorial strength of the major pigment, betanin.

The absorption maximum is in the region of 535 nm when measured in aqueous solution at pH 5.0.

Benefits

Beet red is a water-soluble colour obtained by physical means from a vegetable with a long history of consumption by man. It does not have a specified ADI according to JECFA and is an approved colour for use in foodstuffs according to Directive 94/36/EC. It is listed in Annex V Part 1 as a colour that may be used at *quantum satis*.

Limitations

Betanin is fairly sensitive to heat, light and water activity, and is extremely sensitive to sulphur dioxide. The effect of these limitations may be reduced if specially formulated products are utilised.

Typical Products

Beet red provides a pink hue that is relatively unaffected by pH. It is used in products such as ice cream, dairy products, dessert mixes and icings. Typical dose applications calculated on the basis of the pure pigment are between 5 and 50 ppm.

E163	Anthocyanins
	Grape skin extract
	Grape colour extract
	Enocianina

Sources

Anthocyanins are the mainly red pigments that are responsible for the colours of many edible fruits and berries. They are usually obtained by aqueous extraction, often using sulphurous acid. The major commercial source is grape skins, but anthocyanins are also produced commercially from other edible materials, including elderberries, red cabbage and black carrots.

Function in Food

Anthocyanins are naturally occurring pigments that are widely used to impart either red or purple shades to foods. The appearance of these water-soluble colours is dependent upon the pH of the product in which they are used. Colour hue progresses from red to blue as the pH increases, and the anthocyanins are normally used in acidified products with a pH below 4.5.

The absorption maximum is in the region of 515–535 nm when measured in aqueous solution at pH 3.0.

Benefits

Anthocyanins are water-soluble colours obtained by physical means from edible fruits and vegetables. Accordingly, they have a long history of consumption by man, and normal dietary intake considerably exceeds that used for the purpose of coloration. They do not have an allocated ADI according to JECFA and are approved colours for use in foodstuffs according to Directive 94/36/EC. They are listed in Annex V Part 1 as colours that may be used at *quantum satis*.

Limitations

The stability, shade and colour intensity of anthocyanins are influenced by pH. They are not generally suited for colouring foods with a pH above 4.5.

Some anthocyanins exhibit sensitivity to sulphur dioxide and protein, but these limitations can usually be overcome by careful product selection.

Typical Products

Anthocyanins are particularly useful for the colouring of soft drinks, jams, sugar confectionery and other acidic products, such as fruit toppings and sauces.

Typical dose applications calculated on the basis of the pure pigment are between 10 and 100 ppm.

E170	**Calcium carbonate** **CI pigment white 18**
	Colour Index No: 77220

Sources

Calcium carbonate is a naturally occurring mineral (chalk or limestone), but the food-grade material is made by reaction of calcium hydroxide with carbon dioxide, followed by purification by flotation. As a pigment, the material is also known as CI white 18.

Function in Food

It is used as a colour, a source of carbon dioxide in raising agents, an anti-caking agent, a source of calcium and a texturising agent in chewing gum.

Benefits

Calcium carbonate is readily available and inexpensive. It can be used in raising agents as it releases carbon dioxide both on addition of acid and on heating.

Limitations

Calcium carbonate is a generally permitted additive under Directive 95/2/EC. Calcium carbonate is also permitted as a colour under Directive 94/36/EC. Calcium carbonate is not a bright white colour and titanium dioxide is often preferred.

Typical Products

Calcium carbonate is used in chewing gum and in bread.

E171	Titanium dioxide
	CI pigment white 6

Colour Index No: 77891

Sources
Titanium dioxide is extracted from natural ores and milled to the correct particle size to provide optimum opacity and whiteness. It exists in three different crystalline forms, known as rutile, anatase and brookite. The anatase and rutile form is permitted for use in foodstuffs, however, processing conditions determine the form.

Function in Food
Titanium dioxide is a white powder that is used to colour foodstuffs, to provide opacity and to give a light background for other colourings.

Benefits
Titanium dioxide is the only true white colour permitted in the EU. It is insoluble in water and is stable to heat, light, acids and alkalies.

Limitations
Titanium dioxide is permitted as a colour in a wide range of foodstuffs under Directive 94/36/EC.

Typical Products
Confectionery, ice cream, icings and non-dairy creamers.

E172	Iron oxides and hydroxides

Colour Index No.	Iron oxide yellow: 77492
	Iron oxide red: 77491
	Iron oxide black: 77499

Sources
Iron oxides and hydroxides are produced by the controlled oxidation of iron in the presence of water. The colour range is from yellow through red to black, the precise colour being controlled by the details of the manufacturing process. Even within reds, there are shades from yellow red to blue red.

Function in Food

The iron oxides and hydroxides provide basic yellow, red or black colours to foodstuffs.

Benefits

The pigments are highly stable, particularly to light, and are unaffected by normal food ingredients. They are also suitable for products that are heat-processed.

Limitations

Iron oxides and hydroxides are insoluble in water and give dull colours, which limits their use. Under Directive 94/36/EC, Annex V, Part 1, they are permitted in a wide range of foodstuffs to *quantum satis*.

Typical Products

Fish paste, canned goods, confectionery and pet food.

E173   Aluminium

Colour Index No: 77000

Sources

Food-grade aluminium is 99% pure aluminium.

Function in Food

Aluminium powder is used to colour small decorative pieces used on cakes to give a bright metallic shine.

Limitations

Under Directive 94/36/EC, aluminium is permitted only for the external coating of sugar confectionery for the decoration of cakes and pastries.

E174   **Silver**

Colour Index No: 77820

Sources
Food-grade silver is 99.5% pure silver. Silver is available only as bars or wire. There is no silver-based product analogous to gold leaf.

Function in Food
Silver is used to colour chocolates and liqueurs. Very little silver is used in this way.

Limitations
Silver tarnishes readily in air – powdered silver particularly so.
Under Directive 94/36/EC silver is permitted only for the external coating of confectionery, decoration of chocolates and in liqueurs.

E175   **Gold**

Colour Index No: 77480

Sources
Food-grade gold is 99.99% gold, which is permitted to be mixed with no more than 7% silver or 4% copper. The addition of these metals improves the malleability of the gold so that it can be hammered into gold leaf. Gold mixed with copper has an orange tinge and that mixed with silver a greenish tinge.

Function in Food
Gold is used as gold leaf to wrap chocolate confections, and as tiny pieces to colour confectionery and liqueurs.

Limitations
Under Directive 94/36/EC, gold is permitted only for the external coating of confectionery, decoration of chocolates and in liqueurs.

Typical Products
Chocolate products.

E180	Litholrubine BK
	D & C red no. 6

Sources

Litholrubine is a synthetic azo dye, which is water-insoluble. Its absorption maximum in dimethylformamide is 442 nm.

Function in Food

Litholrubine is a bright red colour used for the edible rind of cheeses. Its major use is in cosmetics.

Limitations

Under Directive 94/36/EC litholrubine BK is permitted only for edible cheese rind.

E200	Sorbic acid
E202	Potassium sorbate
E203	Calcium sorbate

Sources

Sorbic acid is the *trans,trans* isomer of 2,4-hexadienoic acid. It occurs naturally in the unripe fruits of the mountain ash, *Sorbus aucuparia L.* and in some wines. The material of commerce is synthetic.

Function in Food

Sorbic acid and sorbates are used mainly as food preservatives. The antimicrobial activity comprises a wide range of microorganisms, particularly yeasts and moulds, including organisms responsible for mycotoxin formation. Among bacteria, aerobic bacteria are affected the most. Sorbic acid and sorbates are often used in synergistic combination with other preservatives.

Benefits

Sorbic acid and its derivatives can be used in many different types of product featuring a wide range of pH values. They do not interact with other food ingredients, and are neutral in both taste and flavour.

Limitations

Sorbic acid and the sorbates are permitted in Annex III of Directive 95/2/EC as amended by Directives 2003/114/EC and 2006/52/EC in a range of products with individual permitted maxima. The ADI value of sorbic acid and derivatives is 25 mg/kg body weight. Typical usage levels are 2000 ppm in solid food, 300 ppm in beverages. Dry sorbates are very stable. These compounds must not be used in products in whose manufacturing fermentation plays an important role because it inhibits the action of yeast. If potassium sorbate is combined with other preservatives, care must be taken that no calcium ions are present, as this brings about a precipitation. Therefore, for combinations with potassium sorbate, sodium propionate should be used instead of calcium propionate in order to obtain good synergistic action.

Sorbate is believed not to have an effect against oxidative and enzymic browning and should be combined with sulphur dioxide or heat pasteurisation to inhibit browning.

Typical Products

Baked goods, non-alcoholic beverages, cheese, dairy products, delicatessen products, meat products, and fungistatic packing material.

E210	**Benzoic acid**
E211	**Sodium benzoate**
E212	**Potassium benzoate**
E213	**Calcium benzoate**

Sources

Benzoic acid is produced by the oxidation of toluene.

The salts of benzoic acid are made by reacting the acid with the appropriate hydroxide. Sodium benzoate is by far the most common of the three salts used in commerce.

Function in Food

The benzoates are used as preservatives against yeasts and moulds. They have less effect against bacteria. They have been used since the early 1900s.

They are often synergistic with other preservatives, such as sorbates, and are used in conjunction with sulphur dioxide, which itself inhibits enzyme action and browning.

Benefits

The benzoates are readily soluble in water and readily available. Sodium benzoate is the more common product; potassium benzoate is used where a lower sodium content is required. The benzoates are used in acid products, where they are present as benzoic acid.

The acid itself is insoluble in water but moderately soluble in oils.

Limitations

The benzoates have a distinctive flavour, which limits the concentration at which they can be used. They cannot be used in yeast-raised flour products because they inactivate the yeast.

Benzoic acid is only slightly soluble in water.

Benzoic acid and benzoates are permitted under Part A of Annex III of Directive 95/2/EC as amended by Directives 98/72/EC, 2003/114/EC and 2006/52/EC in a range of products, each with its specified maximum concentration.

The ADI is calculated as a total for all benzoates and is 5 mg/kg, expressed as benzoic acid.

Typical Products

Benzoates are the most important preservatives used in soft drinks.

E214	**Ethyl p-hydroxybenzoate**
E215	**Ethyl p-hydroxybenzoate sodium salt**
E218	**Methyl p-hydroxybenzoate**
E219	**Methyl p-hydroxybenzoate sodium salt**

Sources

The esters of p-hydroxybenzoic acid are produced by reacting the respective alcohols with p-hydroxybenzoic acid. The acid itself is made by reaction of potassium phenate with carbon dioxide under pressure at high temperature.

Function in Food

The p-hydroxybenzoate esters are preservatives against yeasts and moulds; they are less effective against bacteria, especially Gram-negative species. Their effectiveness is dependent on the individual species and they are often used in combination or with sorbic or benzoic acid. They are commonly used in cosmetic and personal care products.

Benefits

The esters tend to be more effective antimicrobial agents than benzoic and sorbic acids, and their effect increases with chain length of the ester group. Unlike some other preservatives, they are effective in water at pH from neutral to mildly acid. They are moderately soluble in oils. They tend to be used in combination as their effect is additive but their taste is not.

Limitations

The taste of the esters is detectable even at low levels in food products, so the rate of use is self-limiting. They are only slightly soluble in water. They are permitted in a limited range of snacks, confectionery and pâté, with maximum concentrations defined by Annex III of Directive 95/2/EC.

E220	**Sulphur dioxide**
E221	**Sodium sulphite**
E222	**Sodium bisulphite, sodium hydrogen sulphite**
E223	**Sodium metabisulphite**
E224	**Potassium metabisulphite**
E226	**Calcium sulphite**
E227	**Calcium hydrogen sulphite, calcium bisulphite**
E228	**Potassium bisulphite**

Structure

All the substances that are listed as E220–E228 are equivalent when they are present in food. E221–E228 are all salts of sulphurous acid. This is formed when sulphur dioxide E220 is dissolved in water. The actual species that are present in food depend upon the nature of the food and not upon the chemical form of the additive. It is only in the most acid of foods, e.g. lemon juice and wines, that significant levels of E220 itself occur. Otherwise, the preservative is converted, upon addition to food, into ionic species, mostly hydrogen sulphite and sulphite ion, and into ionic reaction products, all of which are non-volatile. The reason for the relatively large number of "equivalent" substances is technological. Thus, E220 would be used as the additive of choice when fruit is fumigated, or when it is desired to use the substance as an acidulant as well as a preservative. E223 and E224 are particularly stable when stored or handled in the factory environment. On the other hand, E226 is relatively insoluble in water and would be used in situations in which solubility must be minimised. The most widely used form of this preservative is sodium metabisulphite, E223. The term "sulphur dioxide" is used conventionally in the food industry to refer collectively to all these additives, because the recognised methods of analysis convert the additive, in whatever form it is, into sulphur dioxide gas. Legal specifications refer to the mass of sulphur dioxide released upon the analysis of 1 kg of food. However, the individual substances need to be listed with E-numbers because the same mass of each is equivalent to a different amount of sulphur dioxide, and each substance has defined purity criteria. In this section, the term sulphites will be used to refer collectively to substances E220–E228, to avoid confusion with the specific substance, sulphur dioxide.

Sources

All substances in the range E221–E228 are obtained by the addition of sulphur dioxide to the appropriate alkali (sodium, potassium or calcium hydroxide) until the stoichiometric amount has undergone reaction, and the product is then crystallised. Sulphur dioxide is produced synthetically by burning sulphur or, for example, various metal sulphides. It is the starting material for the production of sulphuric acid and so is available cheaply and in a pure state.

Function in Food

Sulphites are the most versatile of all food additives. They have been used in foods since the times of the ancient Romans and Greeks, and are important ingredients in certain traditional foods. Their listing as food preservatives indicates that a primary function is to act as an antimicrobial agent. In this role, sulphites are most effective in acid foods in which the efficacious agent is sulphur dioxide itself. However, sulphites are also added to food to control chemical spoilage, in which capacity they play a unique role. The most well known applications are the control of enzymic browning at the cut or damaged surfaces of plant foods, and non-enzymic browning of sugars or vitamin C when foods are processed thermally or stored. Sulphites inhibit most forms of enzymic spoilage in foods, e.g. those involving oxidising enzymes such as peroxidases and lipoxygenases, which can otherwise cause off-flavours. They prevent oxidative rancidity when unsaturated fats are oxidised non-enzymically in plant foods, and help to preserve vitamins A and C. E223 is used exceptionally as a processing aid to modify the physical characteristics of wheat flour for biscuit manufacture. Sulphites are used to bleach cherries before they are coloured artificially. Contrary to some belief, sulphites do not restore the colour of discoloured meat, but help retain the red colour when used in sausages. The effect of sulphites on any food product is thus seen to be complex and it is recognised that the sensory properties of foods treated with this range of additives differ uniquely from those of the untreated foods. This includes a contribution to the characteristic taste of some foods from sulphur dioxide.

Benefits

As an antimicrobial agent, sulphites are used widely to preserve fermented and non-fermented beverages. Their primary purpose here is to prevent spoilage in storage and after the beverage container has been opened. In wine-making, the resistance of specific yeasts to sulphur dioxide is used to select against "wild" yeasts in fermentation and subsequent storage. It is also thought to contribute to the characteristic dry taste of some white wines. The additive is used against salmonellae and the spoilage yeasts in meat products such as sausage, thus extending the shelf-life of this food. As an anti-browning agent it is used in food production to control enzymic browning after fruits and vegetables are peeled before processing. For this reason, some catering packs of "fresh" peeled potatoes are treated with this additive. It is essential in the production of pale-coloured dried fruit such as apricots, peaches and sultanas. Vegetable dehydration depends critically on sulphites to prevent discoloration during production. Subsequently, sulphites allow dehydrated fruits and vegetables to be stored for long periods of time without specialised storage requirements. In these respects, there are no practical alternatives. As an enzyme inhibitor, sulphites prevent the formation of off-flavours, particularly those that arise from the action of oxidising enzymes on fats. As an anti-oxidant, they help to extend the shelf-life of dehydrated vegetables such as potato and increase the retention of vitamins A and C. They also increase the stability of natural food colours such as the carotenoids (e.g., in dehydrated carrot, peppers, tomato). Sulphites are unique in their control of the staling of beer. As a processing aid, they allow accurate control of the physical properties of wheat flour for biscuit manufacture to ensure a consistent product. A combination of these functions allows fruit to be stored in pulp for many months for jam manufacture without the need for freezing.

Limitations

Sulphur dioxide and the sulphites are permitted under Directive 95/2/EC, as amended by Directives 96/85/EC, 98/72/EC and 2006/52/EC. Sulphites are a normal part of human metabolism even when there is none of the additive in the diet. Whilst the human body is remarkably well able to metabolise and detoxify this additive when ingested, that which is inhaled (as sulphur dioxide gas) can cause an adverse reaction (sometimes severe) in a small number of individuals, particularly those who suffer from asthma. Small concentrations of this gas are present in the headspace above foods in which the additive is present, the highest concentrations being found in acidic food products. There is some concern that individuals who consume large amounts of wine, or have a diet biased towards foods treated with sulphites, can exceed the acceptable daily intake of ingested additive, but there is no known adverse effect arising from such excessive consumption. The classical antinutritional behaviour of the additive is that it destroys vitamin B1 in food, but this is not thought to give rise to vitamin deficiency. A technical limitation is that the amount of the additive present in most foods decreases with time as a result of the many chemical reactions that are required for it to exert its preservative effect. This means that food treated in this way has a limited shelf-life. On the other hand, there are toxicological implications arising from the reaction products. Evidence suggests that a major reaction product formed from sulphite is metabolically inert and, therefore, harmless.

Typical Products

Foods that contain sulphites are too numerous to mention individually but fall into the following classes: soft drinks and fruit juices; fermented drinks, including beer, wine, cider and perry; dehydrated vegetables; dehydrated fruits; peeled potatoes; maraschino cherries; sausages and burgers; jam (as a result of use in fruit pulp); and biscuits. Sulphites may also be present at a very low level in foods where the additive does not serve a technological function, as a result of carry-over in the ingredients or from pre-processing operations.

E234	Nisin

Sources

Nisin is an antimicrobial peptide (small protein) or "bacteriocin" produced by certain strains of the lactic acid bacterium *Lactococcus lactis* subsp. *lactis*.

Commercial preparations, standardised to 2.5% nisin (one million international units per gram), are prepared by the controlled fermentation of nisin-producing *L. lactis* subsp. *lactis* strains in a milk-based medium, recovery and drying of the nisin and blending with salt. Dry nisin preparations are very stable, providing storage is below 25 °C.

Function in Food

Nisin is used as a food preservative and shows strong antimicrobial activity against Gram-positive bacteria but no activity against Gram-negative bacteria, yeasts and moulds. Amongst Gram-positive bacteria, nisin is particularly active against the spore-forming genera *Clostridium* and *Bacillus*. Both spores and vegetative cells are sensitive to nisin, although spores are usually more sensitive than their vegetative cell equivalent. Other non-spore-forming bacteria that are sensitive to nisin are lactic acid bacteria, and *Listeria monocytogenes*.

Benefits

Nisin is an important preservative in foods that are pasteurised but not fully sterilised, since pasteurisation kills Gram-negative bacteria, yeasts and moulds but not bacterial spores. In heat-processed foods, nisin can be used to allow a reduction in the heat-processing regimes, which has the benefit of protecting food against heat damage, thus improving nutritional content, flavour, texture, and appearance, and providing an energy saving.

As an antimicrobial agent against lactic acid bacteria, nisin has applications in low-pH (high-acid) foods such as sauces and salad dressings, and for the control of spoilage lactic acid bacteria in beer, wine and spirit manufacture. In certain foods, such as ricotta, feta and cottage cheese, it can be used to inhibit *Listeria monocytogenes*.

After consumption in food, nisin is degraded by digestive protease enzymes; thus no passage or accumulation of nisin will occur. There are no reported allergic responses of human beings against nisin in food. Nisin is recognised as being of very low or no toxicity and has GRAS (generally recognised as safe) status in the U.S.

Limitations

The Scientific Committee on Food of the European Commission allocated an ADI of 0.13 mg/kg body weight. Usage levels range from 0.5 to 15 mg/kg of food. Under Part C of Annex III of Directive 95/2/EC, nisin is permitted in ripened cheese, processed cheese, clotted cream, mascarpone, semolina, tapioca and similar puddings with individual maximum levels laid down.

Typical Products

Nisin is used in many different foods, e.g. pasteurised processed cheese products, pasteurised dairy desserts, pasteurised milk and milk products, pasteurised liquid egg products, crumpets, canned vegetables, Continental sausages, sauces and salad dressings, and in beer production.

E235   Natamycin (pimaricin, tennectin)

Sources

Natamycin is a natural antimicrobial produced by *Streptomycetes* bacteria found in soil worldwide. Producing organisms are typified by *Streptomyces natalensis*, from Natal, South Africa, where it was isolated in 1955. Commercial preparations are made by the controlled fermentation of dextrose-based media by selected *Streptomycete* strains. Dried natamycin recovered from the fermentation broth is white to cream-coloured, has little or no odour or taste, and in the crystalline form is very stable. Solubility in water and most organic solvents is low.

Function in Food

Natamycin is used as a food preservative. It shows strong activity against yeasts and moulds, but shows no activity against bacteria or viruses. Its effect is predominantly fungicidal.

Benefits

Natamycin is used as a preservative in foods, usually as a surface treatment to prevent the growth of yeasts and moulds. Its low solubility makes it very effective for the surface treatment of foods as it will remain on the surface, and therefore be active at the site where most yeasts and moulds will occur. Natamycin does not interact with other food ingredients, and imparts no off-flavours to food. It is accepted that natamycin is only poorly absorbed from the gastrointestinal tract. Since it is fat- and water-insoluble, the majority of ingested natamycin will be excreted in the faeces. There are no reported allergic reactions to natamycin in food and the product has GRAS (generally recognised as safe) status in the U.S.

Limitations

The ADI of natamycin is 0.25-0.3 mg/kg of body weight/day. Typical usage levels are 2.5 to 8 mg/kg food, primarily as a surface treatment at levels not to exceed 10 $\mu g/cm^2$. Under Directive 95/2/EC, natamycin is permitted in Part C of Annex III for a limited number of uses.

Typical Applications

Natamycin is used as a surface application in cheese coatings, and on dried, cured sausages.

E239	Hexamethylene tetramine

Sources

Hexamethylene tetramine is made by reacting formaldehyde with ammonia. The product is then purified.

Function in Food

Hexamethylene tetramine is a preservative, which works by releasing formaldehyde in acid conditions. The formaldehyde prevents "late blowing" in hard cheese by inhibiting the growth of the bacteria that cause this defect.

Limitations

Hexamethylene tetramine has a slightly sweet taste with a bitter aftertaste.

Under part C to Annex III of Directive 95/2/EC, hexamethylene tetramine is permitted only in Provolone, an Italian hard cheese, to a maximum residual amount of 25 mg/kg, measured as formaldehyde.

E242   Dimethyl dicarbonate (DMDC)

Sources

Dimethyl dicarbonate is not naturally occurring. It is manufactured through chemical synthesis with specially designed extraction and distillation steps to obtain the required purity.

Function in Food

DMDC is used for the cold sterilisation of beverages.

Benefits

Even at very low concentrations, DMDC is very effective against typical beverage-spoiling microorganisms, such as fermentative yeasts, mycoderma and fermentative bacteria. At higher concentrations, it destroys a large number of bacteria, wild yeasts and mould.

Shortly after DMDC has been added to the beverage, it breaks down completely by hydrolysis into minute amounts of microbiologically inactive products.

Sensory tests, and many years of experience have shown that DMDC does not influence the taste, colour or odour of beverages.

Limitations

DMDC is permitted in the EU under Annex III part C of Directive 95/2/EC for use in non-alcoholic flavoured drinks, alcohol-free wine and liquid-tea concentrate, and in the USA, also in wines and their low-alcohol and dealcoholised counterparts.

As the beverage temperature has a major impact on the hydrolysis rate, the beverage should be cooled down to not more than 20 °C, or preferably, below that. This will slow down the decomposition of DMDC, thus prolonging its antimicrobial action. Low temperatures, therefore, support the efficacy of DMDC and its economical use.

Typical Products

Non-alcoholic flavoured drinks, carbonated and non-carbonated juice beverages, ready-to-drink tea beverages, drinks enriched with vitamins, isotonic sport drinks, nutraceutical beverages and juice concentrates.

E249	**Potassium nitrite**
E250	**Sodium nitrite**

Sources

Sodium nitrite is made when a mixture of the oxides of nitrogen is passed into sodium hydroxide. The nitrite crystallises on cooling to a concentrated solution. It is available as both solution and crystals. The major uses of sodium nitrite are in the chemical industry.

There is very little commercial production of potassium nitrite for the food industry.

Function in Food

The nitrites are used as preservatives, preventing the growth of pathogenic microorganisms in meat. They have no action against yeasts or moulds.

By far the majority of nitrite in commerce is the sodium salt.

Benefits

The nitrites are one of the few materials available for preserving cured meats by inhibiting the growth of anaerobic bacteria such as *Clostridium botulinum*. The antimicrobial effect is enhanced when the nitrite is added before the food is heat-processed. They have the added benefits of preserving the red colour (by reacting with the myoglobin) and assisting in the development of the typical "cured" flavour. They also act as antioxidants and prevent the formation of "warmed-over" flavour, which develops when cooked meat is kept exposed to air.

Limitations

The nitrites may be sold for food use only if in a mixture with salt. They are permitted as preservatives only in meat products, sterilised meat products and cured meat products as defined in Annex III part C of Directive 95/2/EC as amended by Directive 2006/52/EC.

The acceptable daily intake (ADI) for the nitrites is given by the Joint FAO/WHO Expert Committee on Food Additives as being between 0-0.06mg/kg body weight.

Typical Products

Sodium nitrite is used in bacon, ham and traditional cured meat products.

| E251 | Sodium nitrate |
| E252 | Potassium nitrate |

Sources

Sodium nitrate is found in nature as Chile saltpetre. It is produced as a by-product of the production of sodium nitrite.

Potassium nitrate is also found as a natural product, but is produced commercially by reacting potassium carbonate with nitric acid.

The major use of potassium nitrate is in agriculture.

Function in Food

The nitrates, particularly the sodium salt, have been used for at least two thousand years as a preservative, often in combination with the nitrite and salt.

They work by being converted into nitrite (see E249/250) in the food by enzymes that are present in the food and in bacteria.

The nitrates are widely present in plant foods and there are traces in water.

Benefits

Use of the nitrates is one of the few methods of inhibiting the growth of anaerobic bacteria, such as *Clostridium botulinum*. This makes them particularly useful in the production of cheese, cured meats and pickled fish.

Limitations

The nitrates are permitted only in cured meats, pickled fish and a number of cheeses as defined by part C of Annex III of Directive 95/2/EC as amended by Directive 2006/52/EC.

Potassium nitrate is less frequently used than sodium nitrate.

The ADI for the nitrate ion is given by the SCF and the Joint FAO/WHO Expert Committee on Food Additives as 0-3.7 mg/kg body weight. This is equivalent to 0-5 mg/kg body weight for sodium nitrate.

Typical Products

Salami, cheese and pickled herring.

E260	Acetic acid

Sources

Vinegar is essentially a solution of 5 to 10% acetic acid in water. The original source of vinegar was the accidental bacterial oxidation of wine, but this has been turned to advantage and today many varieties of vinegar are available derived from different sources of alcohol and flavoured with various herbs.

More concentrated solutions of acetic acid are manufactured industrially by oxidation of ethanol or hydrocarbons.

Function in Food

Acetic acid is naturally present in many foods and has been used for thousands of years as a preservative in pickles. In Rome, it was used in mixtures with salt, wine or honey. It is still used as a preservative, its effect deriving from the decrease in pH. The addition of salt increases its effectiveness, mainly by lowering the water activity.

Benefits

Acetic acid is used as a preservative largely in traditional products, where its flavour contributes to the overall flavour in, for example, pickles, sauces and salad dressings.

Acetic acid is more effective against food-spoilage organisms than would be predicted from the pH and it is effective at higher pH than are other acids. It is synergistic with lactic and sorbic acids.

Limitations

Acetic acid is more effective against yeasts and bacteria than against moulds. In many products it is used in conjunction with other acids, preservatives or preservation methods, such as pasteurisation, to provide additional protection. Its readily recognised flavour limits the applications to savoury products.

It is a generally permitted additive under Directive 95/2/EC.

Typical Products

Acetic acid is used in pickling liquids, marinades, sauces, salad dressings and mayonnaise.

E261   Potassium acetate

Sources

Potassium acetate is made by the reaction of acetic acid and potassium carbonate.

Function in Food

Potassium acetate is used as an acidity regulator and buffer.

Benefits

Potassium acetate is used to modify the flavour of products acidified with acetic acid. Its only advantage over sodium acetate is in products in which the sodium content needs to be reduced.

Limitations

Potassium acetate is a generally permitted food additive under Directive 95/2/EC.

Typical Products

None known.

E262   Sodium acetates
**   (i)   sodium acetate**
**   (ii)   sodium diacetate**

Sources

Sodium acetate is produced by reaction of acetic acid with sodium hydroxide.

When acetic acid and sodium acetate are mixed in equimolar proportions and allowed to crystallise, the sodium acetate crystallises with acetic acid of crystallisation. This material is called sodium diacetate.

Function in Food

Sodium acetate is used as an acidity regulator and buffer.

Sodium diacetate provides a solid source of acetic acid for dry goods.

Benefits

Sodium acetate acts as a buffer and modifies the taste of acetic acid, softening the sharpness of the acid and making it more palatable. It is readily soluble in water.

Sodium diacetate has some specific uses as a source of acetic acid, for example in bread production, where it is used to protect against ropiness and against some moulds, and it is also used as a flavouring in dry products, particularly to impart the flavour of vinegar.

Limitations

The sodium acetates are generally permitted food additives under Directive 95/2/EC.

Sodium diacetate should be stored in well-sealed containers as it loses acetic acid on storage.

Typical Products

Sodium diacetate is used in bread, salt and vinegar flavour snacks, and instant soups.

E263   Calcium acetate

Sources

Calcium acetate is made by the reaction of acetic acid and calcium hydroxide.

Function in Food

Calcium acetate is used as an acidity regulator and as a source of calcium ions.

Benefits

Calcium acetate is readily soluble in water. It can be used to modify the flavour of products acidified with acetic acid or to provide a soluble source of calcium either for fortification or for reaction with alginates.

Calcium acetate is also used to protect against ropiness and against moulds in flour products, but is less effective than sodium diacetate.

Limitations
Calcium acetate is a generally permitted food additive under Directive 95/2/EC. It is more effective against moulds at lower pH levels.

Typical Products
Calcium acetate is used in bread, and in gelling mixtures such as vegetable gelatin.

E270	Lactic acid

Sources
Lactic acid can be produced in a natural and a synthetic manner. The natural form is predominantly present in the L form (the same form as is present in the human body).

Natural lactic acid is commercially produced by fermentation of beet/cane sugar or glucose. No dairy-based lactic acid is commercially sold.

Synthetic lactic acid is produced by a chemical reaction; a racemic mixture of the L and D form is formed.

Function in Food
The main functions of lactic acid are:

Flavouring: lactic acid has a mild, lingering acid flavour

pH-Regulation: lactic acid is used to acidify, because of its mild flavour

Preservation: lactic acid is widely used as a preservative; it inhibits the growth of a wide range of bacteria; because of its mild flavour, relatively high concentrations can be used.

Benefits
Lactic acid is widely used in flavourings, ranging from dairy flavours such as cream, yoghurt and cheese flavours to meat flavours. Further, in mild fruit flavours where other organic acids are too overpowering, lactic acid is used (strawberry, cherry, peach, tropical, etc.).

Lactic acid is used in many applications for its preservative properties. Both spoilage bacteria and foodborne pathogenic bacteria are inhibited by lactic acid. A combination of lactic acid and acetic acid (E260) is used to inhibit the growth of yeasts.

In sugar confectionery products, lactic acid is used to prevent degradation of sugar and gelling agents.

Limitations

Lactic acid is a generally permitted additive under EC Directive 95/2/EC.

Typical Products

Dairy: processed cheese, ricotta, brined cheeses, margarine, spreads
Acidified food products: pickled vegetables, olives, dressings, low-fat mayonnaise, (cooking) sauces, salads
Meat: carcass decontamination
Flavours: component of cheese, cream, yoghurt and meat flavours
Sugar confectionery: soft candies, hard candies (buffered lactic acid)
Bread: sour dough

E280	Propionic acid

Sources

Propionic acid is a saturated fatty acid and a normal constituent of human body fluids. It can be produced by *Propionibacterium* from lactic acid, and also by various methods involving the oxidation of propionaldehyde (a by-product in fuel synthesis and wood distillation).

Propionic acid occurs naturally in ripe Swiss and Jarlsberg cheese at levels as high as 1%; it is also present in the rumen of ruminant mammals. It is digested and metabolised as a fatty acid in humans.

Function in Food

Propionic acid has a preservative effect as a mould inhibitor, and is active against many mould species, including *Aspergillus*, *Penicillium*, *Mucor* and *Rhizopus* – common spoilage organisms in bakery goods. It has a limited inhibitory effect on many yeast species, although it inhibits bakers' yeast species. Its inhibitory effect on bacteria is limited to retarding the growth of *Bacillus subtilis* (rope) in bread.

Benefits

The use of propionic acid extends the mould-free shelf-life of bakery products, cheese and cheese products. It can also prevent blowing of canned frankfurters without affecting their flavour. Propionic acid is more effective at higher pH values, up to pH 6, allowing preservation at higher pH levels than would otherwise be possible. This results in improved sensory quality of food.

Limitations

There are legal limits on the levels of propionic acid in many foods, and it has an ADI of up to 6.0 mg/kg body weight/day expressed as propionic acid.

The use of propionic acid is limited to pre-packed bread and fine bakery wares in the EU according to Annex III Part C of Directive 95/2/EC, although it may be present in certain fermented products resulting from the fermentation process following Good Manufacturing Practice. Propionic acid is permitted under Directive 98/72/EC for surface treatment of cheese and cheese analogues.

It has an optimum activity at pH levels between 5.0 and 6.0 (or higher in some foods). High levels may create bitter, cheesy flavours, and can reduce the activity of bakers' yeast. When using propionates in yeast-leavened products, the yeast level in the formulation should be increased and proof times may need to be extended.

Typical Products

Yeast and chemically leavened bakery products, pre-packed and part-baked bread, cheese and cheese products, pie fillings, tomato purée, canned frankfurters, non-emulsified sauces, artificially sweetened jams, jellies and preserves.

E281	**Sodium propionate**
E282	**Calcium propionate**
E283	**Potassium propionate**

Sources

The propionates are white, free-flowing, water-soluble salts, manufactured by the reaction of propionic acid with carbonates or hydroxides. They are readily digested and metabolised in the body.

Function in Food

The propionates yield the free acid in the product to provide preservative action. They are active against many mould species, including *Aspergillus*, *Penicillium*, *Mucor* and *Rhizopus* – common spoilage organisms in bakery goods. They have a limited inhibitory effect on many yeast species, although they inhibit bakers' yeast species. The inhibitory effect on bacteria is limited to retarding the growth of *Bacillus subtilis* (rope) in bread.

Benefits

The propionates are used to extend the mould-free shelf-life of bakery products, cheese and cheese products. They blend well with other ingredients and do not alter the colour, taste or texture at normal usage levels. Being powders, they are easier to use than liquid propionic acid.

The sodium and potassium salts are recommended for use in chemically leavened products because the calcium in calcium propionate may interfere with some chemical leavening agents.

Limitations

Propionic acid and the propionates are permitted under Annex III Part C of Directive 95/2/EC and Directive 98/72/EC for use in a variety of breads with individual limits, and for surface treatment of cheese and cheese analogues although they may be present in certain fermented products resulting from the fermentation process following Good Manufacturing Practice.

The propionates have an optimum activity at pH levels between 5.0 and 6.0.

High levels may create bitter, cheesy flavours, and can reduce the activity of bakers' yeast. When using propionates in yeast-leavened products, the yeast level in the formulation should be increased and proof times may need to be extended.

Typical Products

Bakery products, including pre-packed bread, Christmas puddings, cakes and pastries and part-baked bread.

| E284 | Boric acid |
| E285 | Sodium tetraborate (borax) |

Sources

Borax is a natural mineral, which is mined and purified. Boric acid is produced by reaction of borax with sulphuric acid followed by purification and crystallisation.

Function in Food

Borax was first recommended as a preservative in 1775. It has been used as a household disinfectant. Both the acid and sodium salt are effective against yeasts, and, to a much lesser extent, against moulds and bacteria.

Benefits

Boric acid has a very low dissociation constant, so it is largely undissociated, and thus effective, even at neutral pH, where carboxylic acid based preservatives have less effect. It is also water-soluble and for many years was used in margarine and butter as it stayed in the aqueous phase.

Limitations

In the opinion of the EU Scientific Committee on Food, boric acid is not suitable for use as a food additive and, under part C to Annex III of Directive 95/2/EC, boric acid and sodium tetraborate are permitted only in caviar to a maximum of 4 g/kg.

E290	Carbon dioxide

Sources

The components of air with approximate ratios are:

78.1%	nitrogen
20.9%	oxygen
0.9%	argon
0.1%	carbon dioxide, rare gases, moisture

Carbon dioxide is normally recovered from flue gases, produced as a by-product of ammonia or hydrogen production, or obtained as an off-gas from fermentation processes. Carbon dioxide is purified and liquefied by a number of different processes.

Function in Food

In modified-atmosphere packaging, carbon dioxide is introduced into a food package to replace air, as an active packaging gas. Carbon dioxide has a powerful inhibitory effect on the growth of bacteria, being particularly effective against Gram-negative spoilage bacteria, such as *Pseudomonas*. The carbon dioxide acts by forming a mild carbonic acid on the surface of the product, lowering the pH and producing an environment relatively hostile to bacteria. The gas can also act as a powerful inhibitor of mould growth.

It is used on its own or in combination with other packaging gases, depending on product and pack format.

Carbon dioxide is also used in the brewing and beverage industries for carbonation of drinks, and is widely used in the food industry for cryogenic chilling, in the form of snow or dry ice.

Benefits

The use of carbon dioxide in packaging extends shelf-life of products, providing benefits in food safety and quality.

Limitations

Carbon dioxide is very easily absorbed into fats and is very soluble in water, particularly as the temperature decreases, so in a retail pack of a food product with high water content there may be sufficient absorption to create a partial vacuum in the pack, causing it to distort or collapse. Under these circumstances, it may be advisable to incorporate a less soluble gas (e.g. nitrogen) into the pack atmosphere along with the carbon dioxide to avoid this collapse, although this may have the effect of reducing the shelf-life of the product.

There have been claims that the use of 100% carbon dioxide can cause a taint to some products, so some food manufacturers specify a reduced level of carbon dioxide in the packaging gas atmosphere.

Carbon dioxide has a relatively high transmission rate through packaging films compared with some other gases, so for the best extended shelf-life the packaging must have good barrier properties.

As it is denser than air and toxic, care should be taken in its use in confined or low-lying working environments.

Carbon dioxide is a generally permitted additive in Annex I of Directive 95/2/EC.

Products

Carbon dioxide is used on its own in extending the shelf-life of bakery products such as par-baked baguettes, hard cheese, bulk poultry in mother bags for storage and distribution, and some fish species. In combination with other gases, it is used to extend the shelf-life of a wide range of products.

E296	Malic acid

Sources

Malic acid occurs naturally in many fruits, including apples, peaches and cherries, but is manufactured industrially from maleic anhydride.

Function in Food

Malic acid is used to provide acidity and to a much lesser extent to chelate metal ions, from hard water or in wine.

Benefits

Malic acid tastes less sharply acid than citric acid and is used to provide an acid taste that is less immediate but persists longer. It is used alone or in combination with other acids to give a range of acid impacts. It is particularly useful in product formulations that use intense sweeteners. It has a lower melting point than citric acid, which is beneficial in the manufacture of boiled sweets.

Limitations

Malic acid is included in Annex I of Directive 95/2/EC as amended–additives that are generally permitted.

Typical Products

Malic acid is used in a wide range of products, including fruit drinks, sports drinks, boiled sweets, chewing gum, sorbets, jams, sweet and sour sauces and peeled potatoes.

E297	Fumaric acid

Sources

Fumaric acid occurs in many plants, but is manufactured by fermentation or by isomerisation of maleic acid.

Function in Food

Fumaric acid provides an acid taste to products.

Benefits

The acid taste of fumaric acid complements and smooths the acidity of other acids.

Fumaric acid is used in powdered products because it is only slightly hygroscopic.

Limitations

Fumaric acid is permitted in Annex IV of Directive 95/2/EC (modified by Directive 98/72/EC) only in fillings and toppings for bakery products, sugar confectionery, some desserts and powdered dessert mixes, chewing gum and instant powders for fruit, tea or herbal tea based drinks, all with specified limits. It is also permitted in some wines. It is poorly soluble in water and dissolves slowly. Fumaric acid has an ADI of 6 mg/kg body weight.

E300	Ascorbic acid (vitamin C)
E301	Sodium ascorbate
E302	Calcium ascorbate

Sources

Ascorbic acid occurs naturally in most fruits and vegetables, notably the citrus family. It is available as an extract of rose hips, but most of the material in commerce is made industrially by a six-step process starting with glucose. Sodium ascorbate is the sodium salt.

Function in Food

In solution ascorbate is easily oxidised to dehydroascorbate, so it is used to prevent oxidation reactions in foodstuffs, extending shelf-life and preserving flavour. The oxidation reaction is readily reversible and some formulations benefit from this.

Benefits

Ascorbic acid is a powerful antioxidant, is synergistic with other antioxidants, and has little flavour. The antioxidant property is used to reduce discoloration in canned fruit and vegetables and in fruit purées caused by polyphenyl oxidase. It is also used in meat products to enhance colour formation and reduce the formation of nitrosamines. Ascorbates are used synergistically with sulphur dioxide in wine and to increase the shelf-life of beer.

Ascorbic acid is used to increase the volume of bread by assisting the formation of the gluten network. It is also added to foods to provide vitamin C as a specific nutrient. It is readily water-soluble.

Sodium ascorbate and calcium ascorbate perform the same functions.

Limitations

Ascorbates are generally permitted for use in foods under Directive 95/2/EC. Ascorbic acid has a slight acid taste and is insoluble in oils; sodium ascorbate has a very slight salty taste. As a raw material, ascorbates are gradually oxidised and should be kept in the dark in sealed containers. Ascorbic acid solutions have a low pH and, where this is a problem, the ascorbates should be used.

Typical Products

Ascorbates are used in an extremely wide range of foods, including bread, canned fruit and vegetables and fruit drinks.

E304	Fatty acid esters of ascorbic acid
	(i) ascorbyl palmitate
	(ii) ascorbyl stearate

Sources

The fatty acid esters of ascorbic acid are made by a two-stage process involving ascorbic acid, sulphuric acid and the individual fatty acids.

Function in Food

The fatty acid esters of ascorbic acid are used to provide the antioxidant capacity of ascorbic acid to oils and fats.

Benefits

The esters are somewhat soluble in fats and oils. The palmitate is more common than the stearate. The reaction with the fatty acid does not affect the antioxidant capacity of ascorbic acid and it still has value to humans as the esters break down in the digestive tract, releasing ascorbic acid.

It is synergistic with dl-α-tocopherol, which is beneficial when stabilising oils with a natural tocopherol content.

Limitations

The ascorbic acid esters are generally permitted for use in foods under Directive 95/2/EC. Because of their limited solubility in oils, they are best dissolved in hot oil, but the temperature needs to be controlled carefully since the palmitate ester, for example, decomposes at 113 °C. As raw materials, the ascorbyl esters are gradually oxidised and should be kept in the dark in sealed containers. The oxidation in solution is catalysed by metal ions, and the protective effect will be limited if metal ions are present in the final product.

Applications

Ascorbyl esters are used in fats and oils, margarines and fat spreads.

E306	Extracts of natural origin rich in tocopherols (natural vitamin E)
E307	Synthetic α-tocopherol (synthetic vitamin E, dl-α-tocopherol)
E308	Synthetic γ-tocopherol (synthetic vitamin E, dl-γ-tocopherol)
E309	Synthetic δ-tocopherol (synthetic vitamin E, dl-δ-tocopherol)

Sources

Tocopherols may either be extracted from vegetable sources – such as oils from soya beans and sunflower seeds, nuts and grains, or produced by chemical synthesis. Although identical in terms of molecular composition, natural and synthetic tocopherols nevertheless exhibit fundamental differences, which are dependent on their origins. For their antioxidant effects in foods, these differences are unimportant, but the tocopherols of natural origin have been shown to be significantly more beneficial in human health and nutrition.

The natural tocopherols each contain only one of the eight possible stereoisomers of the molecule, but the synthetic forms are always a mixture of all eight isomers.

The term "vitamin E" is often used as a general description for the four, closely related, naturally derived tocopherols, all showing some biological activity, but of which the most active and potent is d-alpha-tocopherol. In the naturally derived material, as in the synthetic, the three companion compounds are the beta-, gamma- and delta-isomers. The beta-isomer is present at only about 1% in natural extract and is ignored in the E number nomenclature.The tocopherols are clear, yellow oily liquids, which darken on exposure to light. The tocopherol-rich extract tends to be a darker colour than the individual tocopherols.

Function in Food

The tocopherols are antioxidants. They are fat-soluble and are added to fats and oils to delay or prevent rancidity. It is well to remember that antioxidants cannot reverse or repair damage already done by oxidative processes; nor can they totally prevent it. Only by their presence before, or very soon after the oxidative process begins can they significantly delay the onset of detectable rancidity in fats and oils, which is one of the major causes of the generation of off-flavours in food.

The antioxidant effect of tocopherols and the synthetic antioxidants such as butylated hydroxyanisole (BHA) and butylated hydroxytoluene (BHT) arises from the presence of a hydroxyl group attached to an aromatic ring substituted with methyl groups. This molecular configuration permits donation of a hydrogen atom from the hydroxyl group to a fatty radical, thereby "quenching" it and stopping its catalytic effect in the degradation of oils and fats.

Benefits

The tocopherols also continue their antioxidant role after consumption.

Limitations

The tocopherols are generally permitted additives under Annex I of Directive 95/2/EC.

Typical Products

Margarine and low-fat spreads.

E310	**Propyl gallate**
E311	**Octyl gallate**
E312	**Dodecyl gallate**

Source

The gallates are white, odourless powders prepared by reaction of the appropriate alcohol with gallic acid. They have a slightly bitter taste.

Propyl gallate is also prepared from pods of the Tara tree (*Caesalpinea spinosa*) (see E417) by extraction with propan-1-ol and subsequent purification.

Propyl gallate is the only one of the three gallates in commercial production.

Function in Food

The gallates are antioxidants. They are fat-soluble and are added to fats and oils to delay or prevent rancidity.

Benefits

Propyl gallate is synergistic with other antioxidants, such as BHA and BHT. It is particularly effective with polyunsaturated fats. The antioxidant activity is maintained when the fat is blended with other ingredients in a final foodstuff. The longer chain length of the octyl and dodecyl gallates gives advantages over propyl gallate in terms of greater solubility in fats (and therefore less loss when the fats are emulsified in water) and greater stability.

Limitations

The gallates are included in Part D of Annex III of Directive 95/2/EC as amended by Directive 2006/52/EC, where they are permitted in a range of fats, frying oils and fat-containing products with maximum total limits of gallates, TBHQ, BHA and BHT. If combinations of gallates, TBHQ, BHA and BHT are used, the individual levels must be reduced proportionally.

The gallates are fat-soluble, but need to be dissolved in a small quantity of hot fat first before being diluted with the rest of the fat.

Typical Products

Tallow, polyunsaturated oils.

| E315 | **Erythorbic acid** |
| E316 | **Sodium erythorbate** |

Sources

Erythorbic acid and sodium erythorbate are stereoisomers of ascorbic acid (vitamin C) and sodium ascorbate, respectively. Unlike ascorbates, they do not occur naturally, but are manufactured by a combination of fermentation and organic synthesis.

Function in Food

The erythorbates have the same antioxidant activity as the ascorbates, but with minimal vitamin activity. Therefore, where vitamin activity is not required, erythorbates are used as cost-effective general food antioxidants.

Benefits

Erythorbates are strong reducing (oxygen-accepting) agents, leading to their antioxidant properties in food products in which they are used. Under many conditions, added erythorbates are preferentially oxidised in foods, thus preventing, or minimising, oxidative flavour and colour deterioration, and extending product shelf-life.

Limitations

Under Annex III Part D of Directive 95/2/EC as amended by Directive 2003/114/EC, the erythorbates are limited in their applications to preserved and cured meat up to 500 mg/kg and also in preserved and semi-preserved fish products and frozen and deep frozen fish with red skin, at a maximum use level of 1500 mg/kg (expressed as erythorbic acid). In 1995, the EU Scientific Committee for Food (SCF) allocated erythorbates an Acceptable Daily Intake (ADI) of 6 mg/kg. However, in 1990, the Joint FAO/WHO Expert Committee on Food Additives (JECFA) allocated an ADI of not specified. Consequently, in many other parts of the world, erythorbates are used as general food antioxidants, and may be used at *quantum satis* levels. In the United States, erythorbates have GRAS (generally recognised as safe) status when used in accordance with Good Manufacturing Practice (GMP).

Typical Products
Preserved meat and fish products.

E319	Tertiary butyl hydroquinone (TBHQ)

Sources
TBHQ is an aromatic organic phenol which is chemically synthesised. It is a derivative of hydroquinone, substituted with a tert-butyl group. It is soluble in ethanol but insoluble in water.

Function in Food
Antioxidant

Benefits
TBHQ is rather effective in stabilizing highly unsaturated oils. It offers good carry-through activity to protect fried food products against oxidative deterioration (though it is not effective for baked food applications)

Limitations
TBHQ is permitted for use under Part D of Annex III to Directive 95/2/EC, as amended by Directive 2006/52/EC. It is permitted in a number of foodstuffs all with specified limits. If combinations of gallates, TBHQ, BHA and BHT are used, the individual levels must be reduced proportionally.

The acceptable daily intake (ADI) for TBHQ is given by the Joint FAO/WHO Expert Committee on Food Additives as being between 0-0.7 mg/kg body weight. This ADI was subsequently established by EFSA in 2004.

Typical Products
Vegetable oils for the professional manufacture of heat treated foodstuffs, dry cereals, potato chips, dried meats.

E320	Butylated hydroxyanisole (BHA)

Sources

Butylated hydroxyanisole is a mixture of the 3-tert-butyl and 2-tert-butyl derivatives of 4-hydroxyanisole (also called 4-methoxyphenol). It is produced by the chemical reaction between p-methoxyphenol and isobutene. Preparations usually consist mainly of the preferred 3-tert-butyl isomer.

BHA comprises a white or pale yellowish powder, large crystals or flakes with a waxy appearance and slight aromatic smell. It is soluble in fats, oils, alcohol and ether, but insoluble in water.

Function in Food

Butylated hydroxyanisole is an antioxidant and is added to delay or prevent rancidity in fats and oils in foodstuffs. It is insoluble in water and is best suited to foods with a high fat content. BHA is often used in combination with other antioxidants such as BHT to give a synergistic effect.

Benefits

BHA is stable to heat and mildly alkaline conditions, giving it a property of "carry-through" – a property that makes BHA particularly suitable in baked and fried foods.

Limitations

BHA is a very effective antioxidant for animal fats, but its effect is less marked in vegetable oils that are being stored at ambient temperatures. The acceptable daily intake (ADI) for BHA is given by the Joint FAO/WHO Expert Committee on Food Additives as being between 0 and 0.5 mg/kg body weight per day. Recommended usage rates of BHA are typically 100–200 mg/kg, based on the oil content. BHA is permitted in a limited number of foods under Part D of Annex III of Directive 95/2/EC as amended by Directive 2006/52/EC with individual maxima in each case.

Typical Products

Frying oils, animal fats.

E321	Butylated hydroxytoluene (BHT)

Sources

Butylated hydroxytoluene, 2,6-di-tert-butyl-4-methylphenol, is produced by the chemical reaction between p-cresol and isobutylene.

BHT is a white crystalline solid, either odourless or having a slight aromatic smell. It is soluble in alcohol and ether but insoluble in water.

Function in Food

Butylated hydroxytoluene is an antioxidant and is added to delay or prevent rancidity in fats and oils in foodstuffs. It is insoluble in water and is best suited to foods with a high fat content. BHT is often used in combination with other antioxidants, such as BHA, to give a synergistic effect.

Benefits

The antioxidant activity of BHT can be transferred to baked foodstuffs if it is used as an antioxidant in the shortenings used in their manufacture. The "carry-through" properties of BHT are not as good as those of BHA.

Limitations

BHT is more steam-volatile than BHA, and this makes it unsuitable for use on its own in frying oils, particularly where high-moisture foods are being fried. The acceptable daily intake (ADI) for BHT is given by the Joint FAO/WHO Expert Committee on Food Additives as being between 0 and 0.3 mg/kg body weight per day. Recommended usage rates of BHT are typically 100–200 mg/kg, based on the oil content. BHT is permitted in a limited number of foods under Part D of Annex III of Directive 95/2/EC as amended by Directive 2006/52/EC, with individual maxima in each case.

Typical Products

Tallow, fats and oils.

E322	Lecithin

Sources

Lecithin is a mixture or fraction of phospholipids, which are obtained from animal or vegetable foodstuffs (mainly soya and egg) by physical processes. They also include hydrolysed substances obtained by the use of enzymes. The finished product must not show any residual enzyme activity. A number of different lecithins or lecithin fractions are available.

Function in Food

Phospholipids are the active ingredients of lecithin and have a two-part molecular structure. One part is lipophilic (high affinity to fat/non-polar phase) and the other is hydrophilic (high affinity to water/polar phase). The phospholipids tend to dissolve in fat and disperse in water. This surface activity is the basis for the majority of lecithin applications and allows the formation of both water-in-oil and oil-in-water emulsions.

Besides nutritional benefits, phospholipids have the following functional properties in food products: emulsification and stabilisation of oil-in-water or water-in-oil emulsions; release and anti-spattering effects; adjustment of the flow properties in chocolate masses; improvement of the wettability of instant products; as well as optimisation of the gluten network of baked goods.

Benefits

Lecithin allows the production of fine, stable emulsions with little aggregation or coalescence. It is also used in chocolate manufacture to modify the flow characteristics of liquid chocolate for both blocks and coating. Lecithin is used on the surface of powders to improve "instant" properties. In bakery applications, lecithin is used to increase the extensibility of the gluten in bread making, and in batters to improve the overall distribution of ingredients in cakes and to assist the release of wafers from hot iron moulds.

Limitations

Lecithin is described as a generally permitted food additive in Europe under Directive 95/2/EC.

Typical Products

Margarines, dressings, chocolate and confectionery items, instant powders and bakery goods.

E325	Sodium lactate
E326	Potassium lactate

Sources

Sodium and potassium lactate are produced by neutralisation of lactic acid by sodium hydroxide or potassium hydroxide, respectively. They are available in both a natural and a synthetic form.

Function in Food

The main functions of sodium and potassium lactate in food are: controlling spoilage and pathogenic bacteria; flavouring; and pH regulation where they are used as buffer salts.

Benefits

Both sodium and potassium lactate can be used in pH-neutral food products, such as meat, poultry and fish, and are used at a 2–4% level. Many ingredients become effective only at lower pH, but both sodium and potassium lactate are effective in controlling both spoilage and pathogenic bacteria at neutral pH.

Further, potassium lactate is one of the least bitter-tasting potassium salts available, and can be widely used in the food industry. Both lactates are used as buffer salts in confectionery products, cooking sauces and other savoury flavours.

Both lactates are used to control the fermentation of fermented products such as sausages, and fermented dairy products and vegetables such as pickles.

Limitations

Both lactates have GRAS (generally recognised as safe) status (USA), and can be used to *quantum satis* in most countries, including the EU where they are listed in Annex I of Directive 95/2/EC.

Typical Products

Fresh meat products, sausages, ham, chicken/turkey products, thin sliced food products (deli items), roast beef, convenience food products, whole meal replacements, cooking sauces, confectionery products.

E327   Calcium lactate

Sources
Calcium lactate is produced by neutralisation of lactic acid by calcium hydroxide, chalk or lime. Calcium lactate is made from synthetic D/L-lactic acid or from natural L-lactic acid. The natural form is about twice as soluble as the synthetic form.

Function in Food
Calcium lactate is used as a source of calcium. This is for nutrient fortification, for reaction with pectins in fruit to improve the texture, and to coagulate proteins.

Benefits
Calcium lactate, especially the L form, is very soluble, and is in fact one of the few calcium sources that is soluble in low-pH environments (fruit juice, beverages, pickles, etc.) and neutral-pH environments (milk, diet food, infant food, etc.). Further, calcium L-lactate has a neutral flavour, is highly bio-available and is easily metabolised by the human body.

Limitations
The use of calcium lactate to fortify foods is regulated under EC Regulation 1925/2006. Calcium lactate has GRAS (generally recognised as safe) status (USA), and can be used to *quantum satis* in most countries, including the EU where it is listed in Annex I of Directive 95/2/EC.

Typical Products
Soft drinks, fruit juices, infant food, milk, fruit pastes, pickles, canned fruits, diet foods, sports nutrition products and calcium tablets.

E330   Citric acid

Sources
Citric acid is a key intermediate in the human metabolic cycle. It occurs very widely in nature, most notably in citrus fruits. It was first produced by extraction from lemon juice, but, since the 1920s, it has been made commercially by the large-scale fermentation of sugars using the mould *Aspergillus niger* or yeasts. After a series of purification steps and depending upon the temperature of

crystallisation, either the monohydrate or the anhydrous form is obtained. Citric acid is also produced as an aqueous solution, typically 50% w/w.

Function in Food

Its primary functions in food are as an acid, acidity regulator, antioxidant and sequestrant.

In dilute solution, citric acid reduces the discoloration and spoilage of cut fruits, vegetables and shellfish. It helps prevent rancidity in fats and aids the degumming of vegetable oils.

Benefits

The main characteristic of citric acid is its clean, tart taste, which is compatible with a very wide range of food flavours, both fruit and savoury. As well as giving flavour, its addition to a food formulation lowers pH, which inhibits microbial growth and spoilage. It has antioxidant properties, protecting sensitive flavours, and it is a powerful sequestering agent, binding metal ions that are responsible for the onset of rancidity.

The monohydrate is the "traditional" form; nowadays, most food and beverage formulations are based on the more cost-effective anhydrous form or solution.

Limitations

No ADI is defined for citric acid; unless otherwise specified, it may be used to *quantum satis*. In the EU it is listed in Annex I of Directive 95/2/EC. In practice, its strong acid flavour sets a limit on its use. In foods with an acidic pH, which must be controlled accurately, one should use a buffered mixture of citric acid and citrate. Sodium citrate (E331) is the most widely used, but potassium citrate (E332) may be utilised in low-sodium foods. Citric acid is chemically stable in both its dry form and in solution. It is an irritant, so due care should be taken when handling.

Typical Products

Citric acid is one of the most widely used of all food additives. Its clean, tart taste has found wide application in soft drinks (carbonates, squashes, nectars and powdered beverages), sugar confectionery, jams, jellies, preserves, soups and sauces.

E331	Sodium citrates
	(i) monosodium citrate
	(ii) disodium citrate
	(iii) trisodium citrate

Sources

Citrates are found widely in nature.

The various sodium salts are produced by either partially or completely neutralising citric acid with sodium hydroxide or carbonate. Trisodium citrate dihydrate, which is crystallised with minimum assay 99%, is the most commonly used form.

Function in Food

The sodium citrates are used primarily as acidity regulators, either in combination with citric acid or with acids naturally present in the food formulation. Trisodium citrate is used as an emulsifier in processed cheese and, in combination with ascorbate or erythorbate, it is effective as a cure accelerator in processed meat products.

Benefits

Trisodium citrate is effective and easy to use for pH control in food and beverage products. The appropriate combination of acid and salt can yield pH values from around 6 down to 2. In dry mixes, where moisture content must be minimised, a pH at the upper end of the range can be achieved with trisodium citrate anhydrous and at the lower end with monosodium citrate anhydrous.

Limitations

No ADI has been set; unless otherwise specified, sodium citrate is a generally permitted additive under Directive 95/2/EC and may be used to *quantum satis*. Under Directive 2003/114/EC amending Directive 95/2/EC, it is permitted in UHT goat milk up to 4 g/l.

Typical Products

Trisodium citrate dihydrate, alone, or in combination with citric acid, acts as an acidity regulator in soft drinks, desserts, confectionery, baked goods, preserves and jams.

The other sodium salts, such as trisodium citrate anhydrous and monosodium citrate anhydrous, are usually used only in dry food and beverage formulations.

E332	Potassium citrates
	(i) monopotassium citrate
	(ii) tripotassium citrate

Sources

Citrates are widely found in nature.

The commercially available form of potassium citrate is the tripotassium salt in the monohydrate form. It is produced by neutralising citric acid, usually with potassium hydroxide, followed by crystallisation and drying.

Function in Food

Potassium citrate is used in foods as an acidity regulator and as a source of potassium ion.

Benefits

Potassium citrate offers similar properties to sodium citrate, but it offers greater water solubility and can be used instead of the sodium salt in specifically low-sodium foods. It is also a source of potassium ion in nutritional supplements.

Limitations

No ADI has been set; unless otherwise specified, potassium citrate may be used to *quantum satis* under Directive 95/2/EC. The crystals are very hygroscopic and great care should be taken to prevent them from taking up moisture. When damp, potassium citrate will remain chemically stable but may cake hard. Potassium citrate has a diuretic effect.

Typical Products

Potassium citrate is used in beverages and confectionery.

E333	Calcium citrates
	(i) monocalcium citrate
	(ii) dicalcium citrate
	(iii) tricalcium citrate

Sources

Citrates are found widely in nature. The commercially available form of calcium citrate is the tricalcium salt, formed by completely neutralising citric acid with calcium ion. It is available in powder form.

Function in Food

Calcium citrate is used as an acidity regulator and as a source of calcium ion.

Benefits

Calcium citrate is a physiologically acceptable source of calcium ion. It is effective in the formation of acid-based gels, such as with alginates and pectin.

Limitations

No ADI has been set; unless otherwise specified, calcium citrate may be used to *quantum satis* under Directive 95/2/EC. It has minimal solubility in water, which further decreases with rising temperature. However, a lowering of pH greatly increases solubility.

Typical Products

It is used as a nutrient and dietary supplement in beverages and baby foods; in desserts; and in processed vegetables.

E334	L(+)tartaric acid

Sources

Most of the tartaric acid in commerce is made from the acid potassium tartrate produced as a by-product of the fermentation of grape juice into wine. It is also synthesised from maleic acid.

Function in Food

Tartaric acid has two main uses: to provide a distinctive acid taste to finished products and, as part of baking powder, to react with carbonates to generate carbon dioxide.

Benefits

Tartaric acid has a different taste profile from citric acid, imparting less fresh and more sour notes to products. Naturally, it blends better with grape flavours than with citrus.

Tartaric acid is available as a powder. It is the most water-soluble of the solid acids and is used in baking powders.

It is used as a chelating agent for metal ions naturally present in products, and hence as a synergist for antioxidants.

Limitations

Tartaric acid is generally permitted under Directive 95/2/EC. It has an ADI of 0–30 mg/kg body weight/day (this figure is the total for the acid and its salts).

Typical Products

Baking powder, biscuits and jams.

E335	**Sodium tartrates**
	(i) monosodium tartrate
	(ii) disodium tartrate
E336	**Potassium tartrates**
	(i) monopotassium tartrate
	(ii) dipotassium tartrate
E337	**Sodium potassium tartrate**

Sources

Monopotassium tartrate is formed as a by-product of the fermentation of grape juice. The dipotassium salt is produced by reaction of this with potassium hydroxide.

Sodium tartrates are produced from commercial tartaric acid.

Sodium potassium tartrate is also known as "Rochelle salt", which occurs as a crystalline deposit during the production of wine.

Function in Food

Monopotassium tartrate is also known as "cream of tartar". It is used as a source of acidity in baking powders. Monosodium tartrate would be equally effective but is not so readily available.

The tartrates are also used as buffers and taste modifiers in products containing tartaric acid.

In the past, tartrates were used as emulsifying salts in the production of processed cheese, but they have been replaced because calcium tartrate crystals tended to form during the process, giving the appearance of broken glass in the product.

Benefits

Monopotassium tartrate is one of the fastest reacting acidulants used in baking powder. It also softens the dough but does not weaken it so much that the evolved gas is lost.

Limitations

The tartrates are generally permitted under Annex I of Directive 95/2/EC.

Typical Products

Monopotassium tartrate is used in baking powder.

E338   Phosphoric acid

Sources

Manufactured commercially by the addition of sulphuric acid to phosphate rock, followed by additional steps to remove impurities. Alternatively, formed by burning phosphate rock in an electric or blast furnace to form elementary phosphorus, which is then burned in air to form phosphorous pentoxide. The oxide is hydrated to form phosphoric acid, which is then further purified with hydrogen sulphide.

Function in Food

Phosphoric acid is used as an acidulant, in soft drinks, jams, cheese and beer. It is the only inorganic acid used extensively as a food acid. It is also used as a setting aid and a sequestrant, and in sugar refining.

Benefits

Phosphoric acid is one of the cheapest and strongest food-grade acids available. Its sharp acid flavour particularly complements the dry character of cola drinks – better than citric or tartaric acids. The low pH generated by phosphoric acid is synergistic with other preservatives.

Limitations

Phosphoric acid is included in Annex IV of Directive 95/2/EC as amended, where it is included with the phosphates that are permitted in a wide range of products, with individual limits in each case.

Typical Products

Soft drinks, jam.

E339	Sodium phosphates
	(i) monosodium phosphate
	(ii) disodium phosphate
	(iii) trisodium phosphate
E340	Potassium phosphates
	(i) monopotassium phosphate
	(ii) dipotassium phosphate
	(iii) tripotassium phosphate

Sources

The phosphates are prepared by reaction of metal hydroxides (E524 and E525) with phosphoric acid (E338) under conditions controlled to maximise the yield of the required product. The crystalline phosphates are separated and dried. The products are available both as hydrated crystals and as dehydrated powders. The hydrated form can lose water or cake on storage.

These two sets of products are considered together because they are largely interchangeable, although the potassium salts are generally more soluble than their sodium equivalents.

Function in Food

Monometalphosphates

The monometalphosphates are acidic and can be used as acidulants in raising agents. However, the calcium salt is more commonly used for this purpose (see E341(i) and (ii)).

The sodium and potassium salts are used as chelating agents, buffering agents and occasionally as emulsifying salts in processed cheese products. They are also used as a component of mixtures with other phosphates for protein binding in meat products.

Dimetalphosphates

The dimetalphosphates are used for their ability to enhance water binding in meat and dairy products, preventing water loss and shrinkage during cooking and storage. They are also used to stabilise milk products, such as evaporated milk, against protein coagulation and gelling but also to increase the rate of gelling in instant puddings and cheesecakes. They are powerful sequestrants of calcium in water and are used as such to prevent flocculation of milk proteins during rehydration in hard water of milk-based powders. The dimetalphosphates are the most important emulsifying salts in processed cheese, because they provide the required body and melting performance without fat separation. They are often used in combination with the trimetal phosphates and occasionally with the monometal phosphates.

Trimetalphosphates

Only the trisodium phosphate is of any commercial significance. It is alkaline and is used as a buffering agent and texturiser in meat and cheese products. It is also used to increase the speed of cooking of peas, beans and cereals. The main use is in industrial detergents and toothpastes but, since it is strongly alkaline, it is also used to reduce the microbial load on animal carcasses.

Limitations

Phosphates are included in Annex IV of Directive 95/2/EC as amended, where they are permitted in a wide range of products with individual maximum concentrations.

Typical Products

Processed cheese, cooked ham, desserts, evaporated milk.

E341	Calcium phosphates
	(i) monocalcium phosphate
	(ii) dicalcium phosphate
	(iii) tricalcium phosphate

Sources

The calcium phosphates are manufactured by the reaction of hydrated lime (E526) and phosphoric acid (E338) under conditions controlled to maximise the yield of the required product.

Function in Food

Monocalcium phosphate

Monocalcium phosphate is used as a raising agent when rapid reaction with sodium bicarbonate is required. Unlike the dicalcium salt, the reaction commences as soon as the phosphate is added to the cake batter. Recently, however, mixtures with other phosphates have been developed, which allow a slower rate of reaction, and a more even release of gas.

Monocalcium phosphate is also added to flour to reduce the risk of growth of the bacteria that lead to the spoilage condition known as "rope". It is used as a source of calcium to improve the structure from low-gluten flours, to increase the rate of gelling of some milk-based desserts, and to increase the firmness of canned vegetables such as carrots and tomatoes.

Dicalcium phosphate

Dicalcium phosphate is available in both dihydrate and anhydrous forms. The dihydrate is used as a raising agent in combination with other phosphates and sodium bicarbonate. Dicalcium phosphate is practically insoluble and does not react until the cake is heated to about 60 °C, when it dehydrates and decomposes. It is only mildly acid, having a neutralising value half that of disodium phosphate. It is used in products that require a baking time in excess of 30 minutes and in combination with faster-acting raising agents when it provides last-minute expansion of the cake batter just before the batter sets.

Dicalcium phosphate is also used as a calcium source, to form gels with alginates (E400–403), as a source of minerals in nutrition foods, and as a dispersant in tablets. However, by far the greatest use of dicalcium phosphate is as the abrasive in toothpaste.

Tricalcium phosphate

Tricalcium phosphate is used as a free-flow agent in powdered materials such as icing sugar and powders for instant drinks. Being a fine powder it is used to coat the surfaces of other materials to improve the flowability of the mix and reduce the propensity to form clumps.

Limitations

The phosphates are included with other phosphates in Annex IV of Directive 95/2/EC as amended, where they are permitted in a wide range of products with individual limits in each case.

Typical Products

Monocalcium phosphate is used in cakes, canned fruit and milk desserts. Dicalcium phosphate is used in cakes and tricalcium phosphate is used as a free-flow agent in powdered drinks.

E343	Magnesium phosphates
	(i) monomagnesium phosphate
	(ii) dimagnesium phosphate

Sources

The phosphates are prepared by reaction of magnesium oxide (E530) with phosphoric acid (E338) under conditions controlled to maximise the yield of the required product. The crystalline phosphates are separated and dried. The products are available both as hydrated crystals and as dehydrated powders. The hydrated form can lose water or cake on storage.

Function in Food

Magnesium phosphates are used as acidulants in raising agents in dough.

Benefits

The magnesium phosphates react slowly and are used to stabilise doughs that will be held for some time before baking. They fulfil a similar function to sodium aluminium phosphate.

Limitations

Magnesium phosphates are included with other phosphates in Annex IV of Directive 95/2/EC as amended, where they are permitted in a wide range of products with individual limits in each case.

Typical Products

Bakery goods.

E350	Sodium malates
	(i) sodium malate
	(ii) sodium hydrogen malate
E351	Potassium malate
E352	Calcium malates
	(i) calcium malate
	(ii) calcium hydrogen malate

Sources

The malates are made by reacting malic acid with the appropriate hydroxide or carbonate.

Function in Food

The malates are acidity regulators to buffer and modify the acid taste of products containing malic acid.

Benefits

The malates complement the flavours of products, such as those with apple flavours, acidified with malic acid. The sodium salt is more common than the potassium, which would be used only if the sodium content of the product needed to be restricted.

Limitations

The malates are generally permitted additives under Directive 95/2/EC.

Typical Products

Jam.

E353   Metatartaric acid

Sources

Metatartaric acid is manufactured from glucose. It is also known as glucaric acid.

Function in Food

Metatartaric acid is used as a sequestrant to prevent deposition of cream of tartar (monopotassium tartrate) and calcium tartrate in wine during storage.

Limitations

According to Directive 95/2/EC, metatartaric acid is permitted only in wine (pro memoria) and made wine up to 100 mg/litre.

It is deliquescent and should be kept in tightly closed packages.

E354   Calcium tartrate

Sources

Calcium tartrate is prepared as a by-product of the wine industry.

Function in Food
Calcium tartrate is used as a buffer and as a preservative.

Limitations
Calcium tartrate is a generally permitted additive included in Annex I of Directive 95/2/EC. It is only slightly soluble in water.

Typical Products
None known.

E355   Adipic acid

Sources
Adipic acid is produced by the oxidation of cyclohexane.

Function in Food
Adipic acid is used to provide an acid taste.

Benefits
Adipic acid is used to provide an acid taste with a more lingering flavour profile than citric acid, which works well with some non-citrus fruit products.
It is practically non-hygroscopic.

Limitations
In Directive 95/2/EC, adipic acid is permitted only in fillings for bakery products, dessert mixes and powders for home preparation of drinks, with individual limits specified for each usage. The limits are maxima for any single or combined use of E355, E356 and E357.
The ADI for adipic acid is 5 mg/kg body weight. This covers adipic acid alone or in combination with the sodium or potassium salts of adipic acid.

Typical Products
Individual pies with fruit filling.

| E356 | Sodium adipate |
| E357 | Potassium adipate |

Sources

The adipates are made by reaction of adipic acid with the appropriate hydroxide or carbonate.

Function in Food

The adipates are used to buffer the acidity and modify the acid taste of formulations with adipic acid.

Benefits

The adipates complement the flavour of products acidified with adipic acid. The sodium salt is more common than the potassium, which would be used only if the sodium content of the product needed to be restricted.

Limitations

In Directive 95/2/EC, the adipates are permitted only in fillings for bakery products, dessert mixes and powders for home preparation of drinks, with individual limits specified for each usage. The limits are maxima for any single or combined use of E355, E356 and E357.

The ADI for adipates is 5 mg/kg body weight. This covers the adipates alone or in combination with adipic acid.

Typical Products

Individual pies with fruit filling.

| E363 | Succinic acid |

Sources

Succinic acid occurs naturally in a wide range of vegetables, but is manufactured from acetic, fumaric or maleic acids.

Function in Food

Succinic acid is used to provide a distinctive acid taste.

Benefits

Succinic acid is water-soluble but not hygroscopic, which makes it useful in powdered products.

Limitations

Annex IV of Directive 95/2/EC states that succinic acid is permitted only in desserts, soups and broths and in powders for home preparation of drinks, each with maximum permitted levels. The acid has a pronounced aftertaste and dissolves only slowly in water.

Typical Products

None known.

E380   Triammonium citrate

Sources

Triammonium citrate is the final product of the reaction between citric acid (E330) and ammonium hydroxide (E527). It is a white, water-soluble powder.

Function in Food

Triammonium citrate is little used in the food industry. Its only applications are as a yeast food and a chelating agent.

Limitations

Triammonium citrate is generally permitted under Annex I of Directive 95/2/EC.

E385   Calcium disodium EDTA

Sources

EDTA is ethylene diamine tetraacetic acid.

Calcium disodium EDTA is the mixed salt of EDTA made by reacting the acid with a mixture of calcium and sodium hydroxides. EDTA itself is made by a multistage process starting from ethylene glycol (1,2 dihydroxy ethane).

Function in Food

Calcium disodium EDTA is a sequestrant, both binding metal ions and exchanging its calcium for metal ions.

Benefits

Calcium disodium EDTA is used to sequester small quantities of metal ions present in raw materials or process water. These metals tend to catalyse degradation reactions such as those leading to rancidity, and their removal increases the stability of products during storage and extends shelf-life. By a similar mechanism, it stabilises vitamin C and oil-soluble vitamins.

It is used in spreadable fats as a synergist for the antioxidant vitamins, having the advantage over citric acid or polyphosphate that it imparts no flavour.

The salt is used because it is more stable than the acid.

Limitations

In Directive 95/2/EC, as amended by Directive 2006/52/EC, calcium disodium EDTA is permitted only in a number of canned and bottled products, in spreadable fats and in emulsified sauces with individual maxima specified in each case.

Calcium disodium EDTA has an ADI of 2.5 mg/kg body weight.

Typical Products

Catering sauces and salad dressings.

E400	Alginic acid

Sources

Alginates are the principal structural polysaccharide component of brown seaweeds (just as cellulose is the principal carbohydrate in land plants). The commercial product is extracted from a wide range of brown seaweed species, e.g. Ascophyllum from the North Atlantic, Macrocystis from California and Mexico, Lessonia from South America, Durvillea from Australia, Ecklonia from South Africa, and Laminaria from various northern hemisphere oceans. In general, naturally occurring seaweed is harvested for alginate manufacture, but there is some cultivation in China.

Alginate is present in seaweed as a mixed salt of sodium, potassium, calcium and magnesium. Extraction involves ion exchange in an alkaline medium followed by precipitation, purification and recovery of the alginic acid. Alginic acid is a copolymer of mannuronic acid and guluronic acid – two natural anionic sugars. The monomer composition and sequence vary, mainly as a consequence of the seaweed raw material.

Alginate can also be produced by microbial fermentation, but economics and the need for separate regulatory approval restrict this to a laboratory curiosity at the present time.

Function in Food

Alginic acid swells in water, but does not dissolve, and its main applications are in pharmaceutical tablets. Its swelling ability makes it a useful tablet disintegrant, and it is used in antacid tablets as a raft former for stomach disorders. In the food industry, it is rarely added directly to food compositions. However, it is produced *in situ* when sodium alginate (see E401) is used in acidic foodstuffs. In such situations it will form a gel, skin or fibre as a result of its insolubility in water. Alginic acid is also used in some formulated alginate products for stabilising ice cream and whipped dairy cream. In this case, the alginic acid is converted to sodium alginate *in situ* to provide the stabilisation.

Benefits

Alginic acid and alginates are not absorbed by the human body so are considered a low-calorie ingredient and possibly a source of dietary fibre. They are efficient water binders.

Limitations

Under Annex I of Directive 95/2/EC alginic acid is a generally permitted additive. It is insoluble in water and therefore rarely used directly as a stabiliser or gelling agent. Under Directive 2006/52/EC amending Directive 95/2/EC, E400 is not permitted for use in jelly mini-cups.

Typical Products

Alginic acid is used in ice cream and whipped cream.

E401	Sodium alginate
E402	Potassium alginate
E403	Ammonium alginate
E404	Calcium alginate

Sources

Alginates (see also alginic acid E400) are the principal structural components of brown seaweeds. They are present in seaweed as a mixed salt of sodium, potassium, calcium and magnesium. Extraction involves ion exchange in an alkaline medium followed by precipitation, purification and conversion to the appropriate salt.

Function in Food

The sodium, potassium and ammonium salts are cold-water-soluble, and are used interchangeably, but the calcium salt is insoluble. The salts are used for thickening, gelling, stabilising, film forming and controlled release applications.

Alginates are copolymers of mannuronic and guluronic acids and the monomer composition and sequence vary as a consequence of the seaweed raw material. In general, high guluronic acid alginates are used for gelling applications and the high mannuronic acid types for thickening and stabilising.

The soluble salts form viscous solutions in hot and cold water, and form gels by controlled reaction with calcium. The free calcium content of milk prevents the soluble alginates from dissolving directly in cold milk. This is overcome by the use of calcium sequestering agents or by dissolving milk at, or just below, its boiling point. When a soluble alginate is used as a suspending agent, small amounts of available calcium are beneficial. Any soluble calcium will increase the pseudoplastic nature (shear-dependency) of the alginate solution. At rest, suspended solids, or oil droplets will be stabilised, but the liquid will still flow freely when sheared. Higher concentrations of calcium will produce a thixotropic system (shear-reversible gel) and higher concentrations still will produce a thermostable gel (i.e. it will not melt).

Alginate gels can be internally set, where the gelling ingredients are mixed in with the alginate. Internally set gels are formulated to set within a predetermined time and need to be completely filled into their final container within this time. The careful formulation of partially soluble calcium salts and sequestrants into the product allows the setting time to be varied to fit production needs. Typically, calcium salts such as calcium sulphate and calcium phosphate, and sequestrants such as phosphates and citrates are used for this purpose. Externally set alginate gels rely on the diffusion of readily soluble calcium salts (e.g. calcium chloride, calcium lactate) into food containing an alginate solution. If such a food is extruded into a setting bath containing calcium chloride, a skin of calcium alginate forms instantaneously. This gives the food a structurally robust form and shape. Further calcium diffusion into the centre of the food gels the alginate throughout. In frozen products, e.g. ice cream, sodium alginate prevents ice crystal and fat clump growth during melt/freeze cycles by restricting water mobility.

Benefits

Alginates are very efficient water binders and this leads to their use as thickeners, where low levels give high viscosities; as gelling agents; and in solid foods to prevent water loss, syneresis and phase separation. The cold solubility and the ability to make gels without the use of heat differentiate alginates from other hydrocolloids, e.g. gelatin, agar, carrageenan and locust bean gum, which all require high-temperature processes. This makes alginates particularly useful when used with heat-sensitive ingredients, e.g. flavours, and in applications for safe, convenient domestic use, e.g. instant mousse mixes, cheesecakes. The ability of alginates to form gels, skins and fibres instantly makes them particularly useful for making restructured foods, e.g. onion rings and pet-food chunks. Alginates are not absorbed by the human body, so they are a low-calorie ingredient.

Limitations

As with all hydrocolloids, care needs to be exercised in dissolving alginates. Careless addition leads to clumping, where the outside of the powder hydrates quickly, preventing powder inside the clump from dissolving. The use of appropriate mixing equipment and careful addition of the powder will avoid clumping. Alternatively, the soluble alginates can be dry-blended, e.g. with sugar, or wetted, e.g. with a non-solvent oil or alcohol, prior to addition to water. This will allow each alginate particle to hydrate separately. The soluble alginates will not dissolve directly in cold milk and other high-calcium environments. Sequestrants are normally used to overcome this, and suppliers offer formulated blends. Similarly, the soluble alginates will not hydrate in highly acidic systems (pH <4–5 depending on the grade). Gels are formed by the controlled addition of calcium to alginate solutions. Care and understanding of alginate chemistry are beneficial in deriving optimal functionality. Gels, once formed, are not thermally reversible (they will not melt). Unwanted gelation and rework will need a sequestrant to help recover soluble alginate. The soluble alginates can be used in foods with a pH level as low as 3.5–4.0 (dependent on the grade). Below this, alginic acid precipitates out and propylene glycol alginate (E405) should be considered as an alternative. The soluble alginates are stable at alkaline pH (up to 10, above which depolymerisation is likely). The alginates are generally permitted additives in Annex I of Directive 95/2/EC.

Under Directive 2006/52/EC amending Directive 95/2/EC, E401-E404 are not permitted for use in jelly mini-cups.

Typical Products

Sauces, salad dressings, desserts, fruit preparations, ice cream and water ices, onion rings, low-fat spreads, bakery filling creams, fruit pies, controlled-release pharmaceutical tablets and flavour capsules. Ammonium alginate is particularly used for icings and frostings.

E405   Propylene glycol alginate (PGA)

Sources
        Alginates (see also alginic acid E400) are the principal structural component of brown seaweeds. They are present in seaweed as a mixed salt of sodium, potassium, calcium and magnesium. Extraction involves ion exchange in an alkaline medium followed by precipitation, purification and recovery of the alginic acid. This is then esterified to produce the propylene glycol ester. This will vary in its composition both as a result of the source of the raw material and with regard to its degree of esterification and the percentage of free and neutralised carboxylic acid groups in the molecule.

Function in Food
        Propylene glycol alginate (PGA) is cold-water-soluble, and functions in food as a thickener, suspending agent and stabiliser. It forms viscous solutions in hot and cold water. PGA may be used in many of the same applications as the soluble alginates (E401–3) but has the advantage of being more compatible with more acidic foods and foods with a significant calcium content. This compatibility is a direct result of the esterification of the potentially reactive carboxylic acid groups. The higher the degree of esterification, the higher the compatibility. Conversely, PGA does not form gels or insoluble films and fibres.The residual sensitivity of low- or medium-esterified PGA to calcium can enhance its rheology and provide superior suspending and stabilisation properties, but does prevent it from dissolving in milk below boiling point.
        Grades with a high degree of esterification interact with proteins and are used to stabilise beer foam, meringues and noodles.
        PGA extends the functionality of alginates to lower-pH foods. The soluble alginates are typically not used below pH 4, whereas PGA can be used down to pH 3. It is used in salad dressings to stabilise the oil-in-vinegar emulsion and in fruit drinks to prevent separation of pulp and flavour oils.

Benefits
        PGA is a very efficient water binder, so low levels give high viscosities. In addition to the other alginates it is more acid-stable and less sensitive to calcium. It also interacts with proteins and is particularly useful with heat sensitive systems.

Limitations

As with all hydrocolloids, care needs to be exercised in dissolving propylene glycol alginate. Careless addition leads to clumping, where the outside of the powder hydrates quickly, preventing powder inside the clump from dissolving. The use of appropriate mixing equipment and careful addition of the powder will avoid clumping. Alternatively, PGA can be dry-blended, e.g. with sugar, or wetted, e.g. with a non-solvent oil or alcohol, prior to addition to water. This will allow each alginate particle to hydrate separately. PGA will not dissolve directly in cold milk and other high-calcium environments. Similarly, with acidic foods, optimum functionality is best achieved by dissolving into a neutral system, before adding acid. Propylene glycol alginate can be used in foods with pH values as low as 3.0–3.5 (dependent on the grade). PGA is unstable at alkaline pH and, if protein reactivity at alkaline pH is used in the application, the food product needs to be neutralised quickly after the reaction has occurred. Propylene glycol alginate is permitted in a range of products in Annex IV of Directive 95/2/EC, as amended by Directives 98/72/EC and 2006/52/EC.

Typical Products

Salad dressings, meringues, ice cream, noodles, fermented milk drinks, dairy desserts and beer.

E406   Agar

Sources

Agar is obtained from red seaweeds of the *Gelidium* and *Gracilaria* species collected from the coasts of Japan, Korea, Chile, Spain, Portugal and Morocco, and some is found in Indonesia. Agar is extracted using hot, dilute alkali. The solution is cooled to form a very firm brittle gel, which is frozen to disrupt the gel structure. When the gel is thawed, impurities dissolved in the water can be expelled using high pressure and the gel dried and ground to produce powdered agar. Very small amounts of "natural" strip agar are made from *Gelidium* seaweeds. Solutions are cast in moulds and the gels are frozen naturally, before pressing and drying to give strip agar used in traditional Oriental foods.

Function in Food

Agar forms thermally reversible, firm, brittle gels. These gels are formed by hydrogen bonds between adjacent chains of repeating units of galactose and 3,6 anhydro galactose. This gel structure is not affected by salts or proteins. The gel hysteresis, or difference between melting and setting points, is much greater with agar than with other gelling agents.

Benefits

Agar gels are completely reversible and may be melted and reset without any loss of gel strength. The gels have a characteristic firm brittle texture. Enhanced rupture strength and a more elastic texture are obtained by adding up to 20% locust bean gum, with maximum synergy at 10% locust bean gum, to gels of *Gelidium* agar.

Limitations

Agar is a generally permitted additive in Annex I of EC Directive 95/2/EC. According to Directive 95/2/EC, as amended by Directive 2006/52/EC, E406 is not permitted for use in jelly mini-cups.

In the USA, agar has GRAS (generally recognised as safe) status.

Tannic acid found in some fruits, such as quince and some varieties of apples and plums, can inhibit gelation.

Typical Products

Agar is used in jams and marmalades, toppings and fillings for bakery products such as doughnut glaze. Agar gels are used in gelled meats worldwide. Other applications are largely confined to specific cultural areas of the world. Agar is used to gel fermented dairy products in Europe, but by far the largest volume of agar continues to be used in Asia for traditional dishes of Tokoroten noodles, Mitsumame and Red Bean Jelly.

E407	Carrageenan

Sources

Carrageenans are found in and extracted from certain red seaweeds of the class *Rhodophyceae*. Although some seaweed raw material is gathered from the shores, most is now farmed in such areas as the coasts of the Philippines, Indonesia and east Africa. Carrageenan has a wide range of structures. The polymer chains are based on galactose and anhydrogalactose with varying amounts of natural sulphation.

Function in Food

Carrageenans are used as gelling agents, thickening agents and stabilisers. In dairy products, carrageenans are used to form gels with a range of textures, to thicken milk drinks and to stabilise neutral-pH dairy products. Carrageenan is used to form water jellies, frequently in combination with locust bean gum. The water-gelling properties are widely used in cooked meat products to bind water, especially in cold-eating poultry and pig-meat products. Different seaweed types give different carrageenan types on extraction, which have the designations kappa, iota and lambda. These three idealised types have differences in chemical structure, which lead to differences in gel texture, with kappa types the strongest and most brittle, iota giving soft gels and lambda types being non-gelling. Commercial products frequently are blends of more than one carrageenan type to produce the required textures.

Benefits

Carrageenans can interact with the casein protein in dairy products. This allows carrageenans to produce an equivalent effect in dairy products at lower use levels than most other food gums. The processes used in carrageenan extraction produce gels of good clarity, which is highly desirable in water jellies. Different gel textures are produced by the different seaweed raw materials and so a wide range of textures can be produced in many of the application fields. Carrageenan also has a synergy with locust bean gum, and this is used to extend the range of textures.

Limitations

Carrageenan in solution is not stable to the combination of high temperature and low pH since this will degrade the polymer chain. Carrageenan solutions must therefore be subjected to minimal processing at pH levels below 4.0. Carrageenan is listed in Annex I of Directive 95/2/EC as a generally permitted additive. Under Directive 2006/52/EC amending Directive 95/2/EC, E407 is not permitted for use in jelly mini-cups.

Typical Products

Dairy desserts, powder mixes for dairy desserts, milk drinks, creams and toppings and ice creams; hams and cold-eating poultry products; and glazes for bakery uses.

E407a   Processed eucheuma seaweed (PES)

Sources
Obtained by aqueous alkaline extraction of the red seaweed types *Eucheuma cottonii* and *Eucheuma spinosum* followed by washing, drying and milling.

Function in Food
Processed eucheuma seaweed (PES) is used as a gelling and water-binding agent, and as a thickener and stabiliser. PES is used in hams and cold-eating cooked poultry products to bind water and to increase yields. PES is also used to stabilise ice cream, to thicken flavoured milk drinks, to stabilise cocoa powder in chocolate milks, and to gel dairy desserts.

Benefits
The simpler production process for PES allows a lower-cost product than is possible with carrageenan, and PES can be partially or totally substituted for carrageenan in a number of uses, especially in the meat area. Substantial yield increases can be obtained in cold-eating cooked meat products.

Limitations
PES solutions are not stable to combinations of high temperature and low pH, and must therefore be subjected to minimal heat processing at pH values under 4.0. Insoluble cellulosic components in PES produce cloudy solutions, which are unsatisfactory for many water jelly applications. PES is a generally permitted additive and included in Annex I of Directive 95/2/EC, as amended by Directive 96/85/EC.

Under Directive 2006/52/EC amending Directive 95/2/EC, E407a is not permitted for use in jelly mini-cups.

Typical Products
Hams, poultry roll and chocolate milk.

E410   Locust bean gum (carob gum)

Sources
Locust bean gum (LBG) is the ground endosperm from the seed of the locust bean tree (carob tree), *Ceratonia siliqua*, which grows wild in countries

bordering the Mediterranean Sea. The principal producers are found in Spain, Morocco and Greece. The main component of the white powder (*ca* 80%) is a high-molecular-weight linear polysaccharide (galactomannan) with a mannan backbone chain carrying single galactose residues. The distribution of these galactose sugars along the chain is not known, but statistically there are approximately four mannose sugars present in the molecules for every galactose moiety. In addition to the native gum, LBG is also available in alcohol-washed and alcohol-precipitated qualities. This process removes much of the protein and other components from the gum, which then gives clear transparent solutions. Cold-soluble pregelatinised forms of LBG are also commercially available.

Function in Food

LBG is an efficient thickening and gelling agent. The powder partially hydrates in cold water, but the full viscosity can be obtained only by heating the solution (85 °C). LBG forms thermoreversible gels when mixed with xanthan (ideally in the ratio 1:1). LBG also interacts synergistically with kappa-carrageenans to increase the strength and elasticity of the gels.

Benefits

LBG is used as a thickening agent in hot-prepared fabricated foods. It is widely used in combination with xanthan to prepare elastic gels, which, in comparison with other polysaccharide gelling systems (alginates, carrageenans, pectins) are insensitive to the presence of common cations. Compared with starch, it is more resistant to shear.It is non-digestible and may be classified as soluble fibre. In appropriate dosages, it is known to increase intestinal tract motility and reduce blood serum cholesterol levels.

Limitations

Under Annex I of Directive 95/2/EC as amended, locust bean gum is a generally permitted additive. Because it readily absorbs water and swells, it should not be ingested as a dry powder and, according to Directive 95/2/EC, it is not permitted for use in dehydrated foodstuffs intended to rehydrate on ingestion. Isolated reports have appeared that indicate that the protein in LBG may act as an allergen. The incidence appears to be no higher than that associated with any other natural protein.

Under Directive 2006/52/EC amending Directive 95/2/EC, E410 is not permitted for use in jelly mini-cups.

Typical Products

LBG is widely used as a thickening agent in ice cream and hot-prepared sauces, soups, ketchups and mayonnaises. It is also used for chestnuts in liquid.

It is often found together with xanthan as the gelling system in dressings, desserts and mousses.

E412	**Guar gum**

Sources

Guar gum is the ground endosperm from the seeds of the guar plant (*Cyamopsis tetragonolobus*), which is cultivated in the arid regions of north-western India (Rahjastan) and Pakistan. The main component (*ca* 80%) is a galactomannan with a backbone of mannose to which are attached single galactose residues. The distribution of the galactose along the mannan chain is not known, but statistically there is approximately one galactose residue for every two mannose sugars. The typical molecular weights exceed 106 Dalton but depolymerised grades that show lower viscosity are also commercially available. The powder can be steam-treated to remove much of the characteristic "beany" flavour. Guar from which some of the galactose residues have been enzymically removed so that it mimics the behaviour of locust bean gum towards xanthan and kappa-carrageenan is also on sale.

Function in Food

Guar gum is an efficient thickening agent. It dissolves almost completely in cold water to give opalescent pseudoplastic solutions, and shows a synergistic increase in viscosity when mixed with xanthan. By virtue of its size, guar can cause phase separation with other thermodynamically incompatible solutes. This effect has been exploited in the formulations of fat-reduced or fatless spreads.

Benefits

The pronounced pseudoplastic flow properties of guar solution are ideal for delaying sedimentation of solids or creaming of fats. They ensure that, at low shear forces, an effective viscosity is present without making the product unpalatable.

Guar has an advantage over starch in that it is more resistant to shear.

It is non-digestible and may be classified as soluble fibre. In appropriate dosages, it is known to increase intestinal tract motility and reduce blood serum cholesterol.

Limitations

Guar gum is listed in Annex I of Directive 95/2/EC as amended as a generally permitted additive. Because it readily absorbs water and swells, it should not be ingested as a dry powder and, according to Directive 95/2/EC, it is not permitted for use in dehydrated foodstuffs intended to rehydrate on ingestion. Isolated reports have appeared that indicate that the protein in guar gum may act as an allergen. The incidence appears to be no higher than that associated with any other protein-containing food.

Under Directive 2006/52/EC amending Directive 95/2/EC, E412 is not permitted for use in jelly mini-cups.

Typical Products

Guar gum is widely used as a thickening agent in drinks, sauces, soups, ketchups and mayonnaises. It is also used for chestnuts in liquid. Its cold solubility may be used to advantage in cold-prepared deep-frozen foods. Guar is also widely used as a flour additive in the bakery industry.

E413   Tragacanth

Sources

Tragacanth is a natural gum exudate collected from *Astragalus* species of shrub in response to man-made incisions in the lower stem and root. The exudate is allowed to dry on the shrub prior to collection and is produced in the form of thin white ribbons or larger off-white flakes. The main species are *A. microcephalus* and *A. gummifer*, which grow in arid regions of Iran and Turkey. After collection, the gum is sorted by colour and then milled to a fine powder. Traces of bark and foreign matter are removed before and during the milling process. A wide range of viscosity grades is available, with whiter ribbon grades generally possessing the highest viscosity and the flake form of tragacanth having the best emulsifying properties. Heat-treated variants with lower total viable counts (TVC) are also available. Tragacanth is a complex high-molecular-weight branched polysaccharide consisting of two main fractions. The major fraction (known as bassorin or tragacanthic acid) swells in water and the second fraction (tragacanthin) is water-soluble. Bassorin has a 1-4 linked D-galactose backbone substituted by D-xylose or side chains of D-xylose with L-fucose or D-galactose. Bassorin occurs as a mixed calcium, magnesium and potassium salt. Tragacanthin is a neutral arabinogalactan with a 1-6 and 1-3 linked D-galactose backbone substituted with arabinose side-chains. A proportion of protein (1–4%) is present in tragacanth and may be involved in its emulsifying properties.

Function in Food

Tragacanth is used as a cold-soluble thickener, stabiliser, suspending agent and emulsifier. It is also used as a processing aid in lozenge production, for example, and as a plasticiser in icing. Tragacanth can be used as a fat replacer in emulsion products.

Benefits

Tragacanth is an extremely effective thickener, giving high viscosity at low concentrations. It is unusual in that it possesses both thickening and emulsifying properties; it will thicken and stabilise food emulsions and is particularly effective in pourable emulsions. The excellent acid-stability of tragacanth has resulted in its widespread use in dressings. Tragacanth is also resistant to hydrolysis by food enzymes. Tragacanth possesses suspending properties and has a particularly creamy mouthfeel with neutral flavour. When used to partially replace starch in dressings, for example, there is an improvement in both mouthfeel and flavour. In contrast to stabilisers such as xanthan gum, it does not develop a "stringy" rheology in high-solids systems. Tragacanth can also be made up in concentrated solution (up to 10% w/w), which is an advantage in high-solids systems, where the amount of water is limited (in this instance, hot water may be used to assist hydration). Tragacanth improves the handling and sheeting properties of icing.

Limitations

Tragacanth is relatively expensive and its use in dressings has been largely replaced by xanthan gum. Dispersions of tragacanth can take a long time to hydrate fully unless high-shear mixing is used. Alternatively, warmer water will accelerate viscosity development. Measures may be required to prevent lumping when adding to water. Tragacanth is listed in Annex 1 of Directive 95/2/EC as a generally permitted additive. Under Directive 2006/52/EC amending Directive 95/2/EC, E413 is not permitted for use in jelly mini-cups.

Typical Products

Tragacanth is used in confectionery icing, pourable and spoonable dressings and flavour oil emulsions. Tragacanth has also been used in ice cream and as a suspending agent in fruit drinks and sauces.

E414	Acacia gum (gum arabic)

Sources

Gum arabic is a gummy exudate produced by trees of the species *Acacia senegal* (L.) Willd. and its close relatives as a response to wounding. The majority of the trees are wild, but there are some orchards, mainly in the Sudan. Gum production is encouraged by making a transverse incision in the bark of the trunk and peeling off a thin strip of bark. The gum appears as pale yellow orange tears about the size of a table tennis ball, which harden rapidly by evaporation. The tears are collected by hand and cleaned from loose detritus. Top-quality gum is cleaned finally by dissolving in water followed by filtration and recrystallising or spray drying to produce a powder.

The gum is a polysaccharide with a backbone of D-galactose with D-glucuronic acid units and L-rhamnose or L-arabinose end units.

Function in Food

Gum arabic is used as a viscosity modifier and emulsion stabiliser.

Benefits

Gum arabic is very soluble in water (solutions of up to 50% can be obtained), with a pH of 4.5–5.5. It is practically colourless, odourless and tasteless, and imparts mouthfeel without gumminess. It is good for keeping oils in suspension without a large increase in viscosity, and particularly for encapsulating flavouring oils both for soft drinks and for spray drying to produce powdered flavours. In soft drinks, it allows a long shelf-life and the dried product gives good content to shell ratios and a clean flavour.

Gum arabic can also be regarded as a source of soluble fibre, being unaffected by passage through the stomach but broken down by the large intestine. It is also used to inhibit sugar crystallisation in sweets.

Gum arabic has a property of forming coacervates with gelatin, which forms the basis of its use as a wall material for microencapsulation.

Limitations

Gum arabic is a generally permitted additive according to Annex 1 of Directive 95/2/EC. Under Directive 2006/52/EC amending Directive 95/2/EC, E414 is not permitted for use in jelly mini-cups.

Being a natural product, gum arabic supply is liable to considerable fluctuation and it is increasingly being replaced by modified starches.

Gum arabic is less effective at generating viscosity than most other gums and thickeners.

Typical Products
> Gum arabic is used in soft drinks.

E415	Xanthan gum

Sources
> Xanthan gum is a polysaccharide produced by the fermentation of sugars by the bacterium *Xanthomonas campestris*, which was originally found growing on cabbage leaves. At the end of the fermentation, the broth is sterilised and the gum isolated by precipitation with propanol before washing and drying.

Function in Food
> Xanthan gum is used to increase viscosity in sauces and dressings, drinks and cakes. It is particularly stable to acid, heat and enzymes, meaning that there is no loss in viscosity over the shelf-life of the products.

Benefits
> Solutions of xanthan gum are thick/viscous when at rest but get thinner when they are stirred. The viscosity is regained immediately when stirring stops. This means that they can be used to hold particles in suspension, but the solution will flow easily on stirring or pumping.
>
> In sauces and dressings, xanthan gum is used to provide body and mouthfeel, to increase stability to acid and heat, to provide tolerance to repeated freezing and thawing, and to aid emulsion stability. A useful property is that, when sauces containing xanthan gum are poured out of a bottle, they cut off cleanly and do not drip.
>
> In drinks, xanthan gum is used to improve mouthfeel, particularly in diet products, and to hold particles such as cocoa and orange pulp in suspension.
>
> A major use is in baking, where xanthan is used to reduce splashing during filling moulds, to hold particles such as chocolate chips in suspension while the batter is fluid, and to increase volume in the finished product.

Limitations
> Xanthan gum is a generally permitted additive included in Annex I of Directive 95/2/EC as amended. Because it readily absorbs water and swells, it should not be ingested as dry powder and, according to Directive 95/2/EC, it is not permitted for use in dehydrated foodstuffs intended to rehydrate on ingestion. Therefore it cannot be used to produce dehydrated foods intended to rehydrate

upon ingestion. Its rapid rate of hydration means that it is important to ensure that it is well dispersed throughout a mix before water is added, or it can form lumps. Under Directive 2006/52/EC amending Directive 95/2/EC, E415 is not permitted for use in jelly mini-cups.

Typical Products

Sauces and dressings, drinks, cakes, fruit preparations, desserts, meat products and chestnuts in liquid.

E416	Karaya gum

Sources

Karaya gum is a natural tree exudate collected from *Sterculia urens* (Roxburgh) and other species of *Sterculia* and *Cochlospermum* following man-made incisions in the bark of the tree. The main growing areas are India and West Africa (Senegal and Mali). The gum is allowed to dry before collecting, after which it is sorted by colour and then milled to a powder. Traces of bark and foreign matter are removed before and during the milling process. Different grades are classified by colour, particle size and viscosity. West African material generally possesses a more pseudoplastic rheology than material from India. Karaya gum is a high-molecular-weight, branched anionic polysaccharide, which occurs as a partially acetylated, mixed calcium and magnesium salt. The structure of karaya gum is not fully understood, but it appears to consist of a backbone based on D-galacturonic acid and L-rhamnose with side chains of D-galactose and D-glucuronic acid. The ratio of these constituents varies depending on the source of karaya. Overall, karaya contains approximately 37% uronic acid residues and 8% acetyl groups.

Function in Food

Karaya gum is used as a thickener (cold make-up) and as a coating and glazing agent. Karaya particles do not normally dissolve but swell in a similar fashion to starch, although karaya generally thickens at a lower concentration than starch, forming a thick paste at 3% w/w in water.

Benefits

Karaya gum has good acid-stability and is resistant to hydrolysis by food enzymes. Karaya is useful as a thickener since it does not have the "gummy" texture associated with many other hydrocolloids. It can also provide a better flavour release than the equivalent level of starch. The texture of a karaya paste in water can to some extent be controlled by the original particle size of the dry powder. Karaya gum has adhesive properties. It is resistant to human digestive enzymes and therefore has a low caloric value. Its indigestibility has resulted in its use as a laxative.

Limitations

The acidic flavour of karaya gum has limited the number of applications for the gum. In order to achieve maximum viscosity, karaya gum should be dispersed into water prior to the addition of other ingredients such as acid or sugar. Measures may be required to prevent lumping when adding to water. Karaya gum is not normally used at a pH higher than 7 since its rheology changes to a ropey mucilage as a result of deacetylation. The stability of gum karaya in powder form (with respect to water viscosity) is not as good as that of some hydrocolloids. Karaya gum is placed in Annex IV of Directive 95/2/EC, as last amended by Directive 2006/52/EC, which limits its applications to a number of products, with specific maxima in each case.

Typical Products

One of the main food uses for karaya gum is as a thickener in sauces, in particular brown sauce. It is also used in coatings, fillings, toppings and chewing gum.

E417	Tara gum

Sources

Tara gum is the ground endosperm from the seed of the Tara shrub *Caesalpinia spinosa*, which is endogenous to Peru and Ecuador. The main component of the powder (*ca* 80%) is a high-molecular-weight linear polysaccharide (galactomannan) with a backbone chain of 1-4 linked β-D-mannose residues, to which 1-6 α-D-galactose sugars are attached. The distribution of the galactose along the mannan chain is not known, but statistically there is approximately one galactose residue for every three mannose residues.

Function in Food

Tara gum is an efficient thickening and gelling agent. It dissolves partially in cold water, generating *ca* 70% of its potential functionality. It hydrates fully in water above 85 °C, forming an opalescent pseudoplastic solution. Mixed with xanthan, it forms thermoreversible gels and increases the elasticity of kappa-carrageenan gels.

Benefits

Tara gum can be used as a thickening agent in fabricated foods. It is more resistant to shear than starch.

It is non-digestible and may be classified as soluble fibre. In appropriate dosages, it is known to increase intestinal tract motility and reduce blood serum cholesterol levels.

Limitations

Tara gum is a generally permitted additive in Annex I of Directive 95/2/EC. Because it readily absorbs water and swells, it should not be ingested as a dry powder and, according to Directive 95/2/EC, it is not permitted for use in dehydrated foodstuffs intended to rehydrate on ingestion.

Under Directive 2006/52/EC amending Directive 95/2/EC, E417 is not permitted for use in jelly mini-cups.

Typical Products

Tara gum may be used as a thickening agent in sauces, soups, ketchups and mayonnaises.

In the majority of European countries, the use of tara gum in foods was not permitted before the miscellaneous Directive came into force in September 1996. Therefore, only a limited number of applications for this gum have been found.

E418   Gellan gum

Sources

Gellan gum is an extracellular polysaccharide secreted by the microorganism *Sphingomonas elodea*, previously referred to as *Pseudomonas elodea*.

Gellan gum is manufactured by inoculating a fermentation medium with the microorganism. The fermentation is carried out under sterile conditions with strict control of aeration, agitation, temperature and pH. After fermentation, the viscous broth is pasteurised to kill viable cells. The polysaccharide can then be recovered in several ways. Direct recovery by alcohol precipitation from the broth yields the substituted, high-acyl form. Alternatively, treatment of the broth with alkali prior to alcohol precipitation results in deacylation and yields the unsubstituted, low-acyl form.

Gellan gum is a mixed salt, predominantly in the potassium form, but also containing other ions such as sodium, calcium and magnesium.

Function in Food

Gellan gum is soluble in hot water and functions in food as a gelling, stabilising, film-forming and suspending agent. The properties of gellan gum are dependent on the degree of acyl substitution. Gellan gum forms gels when hot solutions are cooled in the presence of gel-promoting cations such as sodium, potassium, calcium or magnesium. Gelation and hydration of the high-acyl form are less dependent on ions than is the case with the low-acyl form. Calcium in particular inhibits hydration of the low-acyl form. This can be overcome by the use of sequestering agents such as phosphates or citrates, which are commonly used in foods. Both forms of gellan gum will hydrate in hot milk without the need for a sequestrant. The precise setting temperature of the gels will depend on the type of gellan gum, which cations are present and their concentration, and the presence of other dissolved solids. Low-acyl gellan gum forms gels at temperatures typically between 30 and 50 °C, while high-acyl gellan gum normally forms gels at around 70 °C. To obtain optimum gel properties it is sometimes necessary to add extra cations, usually a soluble calcium salt. Any addition of cations is best carried out while the solution is hot. Texturally, the low-acyl form produces firm brittle gels, whereas the high-acyl form produces soft elastic gels. Intermediate textures can be produced through mixtures of the two forms. At very low concentrations, gellan gum gels can be sheared to produce smooth, homogeneous, pourable structured liquids, sometimes referred to as fluid gels. These systems exhibit a yield stress (weak gel structure), which enables effective suspension of materials such as jelly beads, herbs or spices. These fluid gels have very low viscosity when poured or drunk, and are ideal for suspension of particulate materials in beverages.

Benefits

Gellan gum is effective at very low concentrations and does not mask the flavours in the food. Using the two forms of the gum in combination allows a wide range of textures to be produced. Gellan gum gels can be formed over a wide range of pH levels, from 2.5 to 10.0, and a wide range of soluble solids, from 0 to 75% total soluble solids.

Limitations

Gellan gum is a generally permitted additive in Annex I of Directive 95/2/EC. Combinations of high-acyl and low-acyl gum are both included under the one designation of gellan gum. Under Directive 2006/52/EC amending Directive 95/2/EC, E418 is not permitted for use in jelly mini-cups.

As with all hydrocolloids, care needs to be exercised in hydrating gellan gum. Careless addition leads to clumping, where the outside of the powder hydrates quickly, preventing powder inside the clump from dissolving. The use of appropriate mixing equipment and careful addition of the powder will avoid clumping. Alternatively, gellan gum can be dry- blended, e.g. with sugar, or wetted, e.g. with a non-solvent oil or alcohol, prior to addition to water. This will allow each gellan particle to hydrate separately. Care and understanding of gellan gum chemistry are beneficial in deriving optimal functionality. Low-acyl gellan gum gels are generally not thermally reversible (they will not melt) and therefore rework to recover the solution is difficult. Gellan gum, like all hydrocolloids, will degrade in hot acidic conditions; therefore, it is recommended to add any acid at the final stage of gel preparation just prior to filling and cooling of the system.

Typical Products

Gellan gum is used in a wide range of fruit preparations, including fruit fillings and high-solids bakery jams. As a suspending agent, it is used in beverages to suspend fruit pulp or jelly beads. It is also used in combination with gelatin to improve the thermal stability of aspics. Gellan gum is used to produce gummy confectionery for the Kosher/Parve market. It is also used to aid the adhesion of salts and flavours to the surface of crisps and nuts.

E420	**Sorbitol**
	(i) crystalline sorbitol
	(ii) sorbitol syrup

Sources

Sorbitol is widely present in nature, particularly as a constituent of many fruits and berries. Commercial products are manufactured by hydrogenation of dextrose and dextrose/glucose syrup, followed, for the crystalline sorbitol, by crystallisation.

Function in Food

Sorbitol exists as a pure crystalline material and as aqueous solutions having a dry matter content of 70%. Sorbitol is a nutritive sweetener and replaces sucrose and glucose syrups, for bulk, texture and sweetness, in sugar-free confectionery products such as chewing gum, compressed tablets and hard-boiled, soft and chewy candies. Sorbitol syrup is also used as an efficient humectant, and as a sequestering and emulsifying agent in confectionery and bakery products, as well as in mayonnaise, creams and sauces.

Benefits
Sorbitol does not promote tooth decay and has a reduced caloric value (2.4 kcal/g in Europe, 2.6 kcal/g in the USA). It extends the shelf-life of food products and does not provide browning in food when heated or baked. Sorbitol can be combined with other polyols as well as with intense sweeteners to balance its slightly reduced sweetening power (*ca* 60% of sucrose). Sorbitol is well tolerated by diabetics.

Limitations
Sorbitol is widely permitted *quantum satis* in both the sweeteners Directive 94/35/EC as amended and Annex IV of the miscellaneous additives Directive 95/2/EC for use in foods in general for purposes other than sweetening. As with all the polyols and some sources of dietary fibres, excessive consumption of sorbitol can produce a laxative effect.

Typical Products
Sugar-free confectionery products such as chewing gum, compressed tablets, hard-boiled, soft and chewy candies and chocolate; bakery products; and fish and surimi products.

E421   Mannitol

Sources
Mannitol is widely present in nature, particularly in fruits, plants and algae. Commercial mannitol is manufactured by hydrogenation of fructose or mannose, followed by a crystallisation and drying step.

Function in Food
Mannitol exists as a pure crystalline material. It is a nutritive sweetener and specifically controls water activity to help reduce stickiness, particularly in chewing gum and hard-boiled candies. Mannitol is 50–60% as sweet as sucrose.

Benefits
Mannitol does not promote tooth decay and has a reduced caloric value (2.4 kcal/g in Europe, 1.6 kcal/g in the USA). It extends the shelf-life of food products and does not provide browning in food when heated or baked. Mannitol can be combined with other polyols as well as with intense sweeteners. It is well tolerated by diabetics.

Limitations

Mannitol is widely permitted *quantum satis* in both the sweeteners Directive 94/35/EC as amended and Annex IV of the miscellaneous additives Directive 95/2/EC for use in foods in general for purposes other than sweetening. As with all the polyols and some sources of dietary fibres, excessive consumption of mannitol can produce a laxative effect.

Typical Products

Sugar-free chewing gum, sugar-free hard-boiled candies, and chocolate.

E422   Glycerol

Sources

Glycerol is made by the hydrolysis of fats. It can be obtained from both animal and vegetable fats, and material from both sources is readily available.

Function in Food

Glycerol is liquid at room temperature. It is used as a humectant, to keep foodstuffs moist to the palate without the risk of mould or bacterial growth. It is also used to retard staling and to improve texture by plasticising the food.

Benefits

Glycerol is naturally present in food and is formed in the human digestive system. It is readily available and has a long history of use.

It is liquid at room temperature and adds moistness to products, at the same time decreasing water activity. Moistness could also be produced by using more water, which introduces the risk of mould growth, or by using sugar syrup, which increases sweetness. Glycerol avoids the former problem with only a slight increase in sweetness. In products where sugar crystallises after manufacture, glycerol is used to inhibit crystallisation, thus maintaining more sugar in solution, which itself has a humectant effect.

Glycerol is also less volatile than water, which means that it is better at maintaining moistness over the shelf-life of the product.

Limitations

Glycerol is generally permitted under Annex I of Directive 95/2/EC. It does have the particular taste effect of leaving a slight burning sensation in the throat, which limits the quantities that can be used in a product.

Typical Products
Glycerol is used in cakes and confectionery.

E425	Konjac
	(i) konjac gum
	(ii) konjac glucomannan

Sources
Konjac gum and konjac glucomannan, also known as konjac flour, yam flour, konnyaku glucomannan and glucomannan gum, are extracted from the tuber of the *Amorphophallus konjac* plant. Tubers are harvested after 2–3 years, when they contain 30–50% glucomannan, which is sufficient for commercial extraction. After harvesting, the tubers are washed and cleaned quickly to avoid bruising and spoilage, followed by slicing and chipping to assist drying. The dried tubers are ground and separated by air classification. The heavier idioblast sacs, which contain the konjac gum, are recovered and washed with alcohol and water to remove starch, protein and other unwanted materials together with the strong fishy taints naturally associated with konjac. Finally, the powder is dried, ground and blended.

Function in Food
Konnyaku noodles are a traditional food in the Far East, made by heating glucomannan solutions with limewater to form a thermally stable gel, which is cut into thin strips and used as a meal component. In table dessert gels and aspics, 0.6% of a konjac gum-kappa-carrageenan blend gives firm cohesive textures. The thermally stable glucomannan gel is used in coarse-ground sausage and meat analogues as a texture modifier and water binder. Konjac gum acts as a binder and protects against freezer damage in surimi. In cream cheese and processed cheese, a low level of around 0.2% glucomannan is very effective for moisture binding and good spreading properties and for giving a creamy mouthfeel and full body. Konjac gum provides ice crystal control, thickening and bodying to ice cream and frozen desserts. It is used in sauces, gravies, salad dressings and mayonnaise for thickening and stabilising. In bakery applications, the glucomannan acts as a film former and flow aid for coatings, toppings and batters and as a binder and extrusion aid for pasta.

Benefits
The high molecular weight of around 1,000,000 Daltons for this linear glucomannan confers a high viscosity when the gum is fully hydrated. The non-

ionic D-mannose and D-glucose units in konjac gum are relatively unaffected by high levels of salt, and the glucomannan is stable to below pH 3.8. The gum is a source of soluble fibre as the β1-4 linkages in the glucomannan chain resist enzymic degradation during digestion.

The glucomannan contains random acetyl side groups, which prevent long-chain polymers from associating to form a gel. The acetyl groups can be removed by adding a weak base to raise the konjac solution above pH 9, and heating. Once the side groups are removed, the polymer chains interact to form non-melting gels. The rate of gel formation is controlled by pH and temperature. Gelation proceeds as the gel is deacetylated so that a gel may be formed at any temperature: there is no specific setting temperature as in the case of carrageenan, agar or gelatin. Gels are insoluble in water and are stable to temperatures above 200 °C.

Adding 0.02 to 0.03% konjac flour to 1.0% xanthan gum will increase the viscosity two or three times through inter-chain associations between the two polymers. Higher levels of konjac will form a thermally reversible gel with xanthan. Blends of konjac and kappa-carrageenan show stronger synergy than blends of carrageenan and locust bean gum. Heat is required to hydrate both gums fully and a thermally reversible gel forms upon cooling. By varying the gum ratio, the texture can be varied.

Limitations

Konjac has been consumed in foods for over 1,000 years in Asia, and is considered a food product. To the rest of the world, it is a relatively new food ingredient. In the EU, Directive 95/2/EC as amended by Directive 98/72/EC and further amended by Directive 2003/52/EC permits levels up to 10 g/kg of konjac gum and konjac glucomannan for general food use however, it may not be used to produce dehydrated foodstuffs intended to rehydrate on ingestion, nor in jelly confectionery including jelly mini-cups. Konjac gum hydrates in water at room temperature, but heating or shearing the solution greatly speeds up this process.

Typical Products

Aspics, surimi, frozen desserts, sauces and batters.

E426	Soybean hemicellulose

Sources

Soybean hemicellulose is extracted from soya fibre. Soya fibre is a mixture of cellulosic and noncellulosic structural components of the internal cell wall of soya beans. Its major fractions are noncellulosic. The raw material ('Okara') from which the soluble hemicellulose is extracted, is a high-fibre containing by-product of the soya oil and soya protein production process.

Function in Food
Emulsifier, thickener, stabiliser, anti-caking agent.

Benefits
Soybean hemicellulose is unlikely to degrade. The presence of water-soluble hemicellulose in the diet doesn't affect the bioavailability of other nutrients.

Limitations
Soybean hemicellulose is permitted under Annex IV to Directive 95/2/EC, as amended by Directive 2006/52/EC. It is allowed in dairy based drinks, food supplements and emulsified sauces. It is also permitted in a number of pre-packaged products: fine bakery wares, ready to eat oriental noodles and rice, processed potato and rice products, processed egg products, jelly confectionery (but excluding jelly mini-cups), all with specified limits. Hemicellulose derived from soybean presents problems for allergy sufferers.

Typical Products
Dairy based drinks; emulsified sauces; oriental noodles and rice products; jelly confectionery except jelly mini-cups.

| **E431** | **Polyoxyethylene (40) stearate** |

Source
Polyoxyethylene stearate is made by reacting stearic acid with polyoxyethylene, a polymer of ethylene oxide.

Limitations
In Directive 95/2/EC, polyoxyethylene stearate is permitted only in wine that has been imported from certain countries, where it is used to inhibit foam formation during fermentation. It is not permitted in wine made within the EU.

E432   Polyoxyethylene sorbitan monolaurate (Polysorbate 20)

Sources

Polysorbate 20 is a pale yellow liquid produced from a mixture of partial laurate esters of sorbitol and its anhydrides, condensed with ethylene oxide.

Function in Food

Polysorbate 20 is widely used within the food industry as a surfactant, for forming oil-in-water emulsions such as dressings, sauces and margarines. The surfactant properties also lead to uses in improving the volume and texture of cakes, the dispersion of coffee whiteners, and the aeration, dryness and texture of whipped cream.

Benefits

Polysorbate 20 can be used in combination with mono- and diglycerides of fatty acids (E471) or other polysorbates to provide the optimum balance of emulsion properties. It is soluble in hot and cold water but insoluble in edible oils.

Limitations

Polysorbate 20 is one of the polysorbates included in Annex IV of Directive 95/2/EC as amended by Directives 2003/114/EC and 2006/52/EC, where they are permitted in a range of product categories, with limits in each case. Polysorbate 20 has a warm, somewhat bitter taste.

Typical Products

Cakes and cake mixes, coffee whiteners, whipped creams based on dairy and vegetable fat, margarine, salad dressings and sauces.

E433   Polyoxyethylene sorbitan monooleate (Polysorbate 80)

Sources

Polysorbate 80 is a pale yellow liquid produced from a mixture of partial oleate esters of sorbitol and its anhydrides, condensed with ethylene oxide.

Function in Food

Polysorbate 80 is used as a surfactant, often in combination with other emulsifiers for forming oil-in-water emulsions. It is used to stabilise margarine, sauces and dressings, and to hold the fat in ice cream.

Benefits

Polysorbate 80 can be blended with other polysorbates or mono- and diglycerides of fatty acids (E471) to provide the optimum balance of emulsion properties. It is soluble in hot and cold water but insoluble in edible oils.

Limitations

Polysorbate 80 is one of the polysorbates included in Annex IV of Directive 95/2/EC, as amended by Directives 2003/114/EC and 2006/52/EC where they are permitted in a range of product categories, with limits in each case. Polysorbate 80 has a warm, somewhat bitter taste.

Typical Products

Ice cream, frozen desserts, margarine, salad dressings and sauces.

E434   Polyoxyethylene sorbitan monopalmitate (Polysorbate 40)

Sources

Polysorbate 40 is a pale yellow liquid produced from a mixture of partial palmitate esters of sorbitol and its anhydrides, condensed with ethylene oxide.

Function in Food

Polysorbate 40 is widely used as a surfactant, often in combination with other emulsifiers for forming oil-in-water emulsions. It is used to stabilise sauces and dressings and in bakery margarine to improve aeration, cake volume and texture. It is also used in whipped cream and coffee whiteners.

Benefits

Polysorbate 40 can be blended with other polysorbates or mono- and diglycerides (E471) to provide the optimum balance of emulsion properties. It is soluble in hot and cold water but insoluble in edible oils.

Limitations

Polysorbate 40 is one of the polysorbates included in Annex IV of Directive 95/2/EC as amended by Directives 2003/114/EC and 2006/52/EC, where they are permitted in a range of product categories, with limits in each case. Polysorbate 40 has a warm, somewhat bitter taste.

Typical Products

Cakes and cake mixes, coffee whiteners, whipped creams based on dairy and vegetable fat, margarine, salad dressings and sauces.

E435   Polyoxyethylene sorbitan monostearate (Polysorbate 60)

Sources

Polysorbate 60 is a pale yellow liquid or semi-gel at room temperature. It is produced from a mixture of partial stearate esters of sorbitol and its anhydrides, condensed with ethylene oxide.

Function in Food

Polysorbate 60 is used as a surfactant, often in combination with other emulsifiers for forming oil-in-water emulsions. It is used in bakery margarine to improve dough conditioning and reduce staling in bread and to improve batter aeration in cakes. It is also used for dressings and sauces.

Benefits

Polysorbate 60 can be blended with other polysorbates or mono- and diglycerides of fatty acids (E471) to provide the optimum balance of emulsion properties. It is soluble in hot and cold water but insoluble in edible oils.

Limitations

Polysorbate 60 is one of the polysorbates included in Annex IV of Directive 95/2/EC as amended by Directives 2003/114/EC and 2006/52/EC, where they are permitted in a range of product categories, with limits in each case. Polysorbate 60 has a warm, somewhat bitter taste.

Typical Products

Cakes and cake mixes, coffee whiteners, whipped creams based on dairy and vegetable fat, margarine, salad dressings and sauces.

E436   Polyoxyethylene sorbitan tristearate (Polysorbate 65)

Sources

Polysorbate 65 is a tan colour solid produced from a mixture of partial stearate esters of sorbitol and its anhydrides, condensed with ethylene oxide.

Function in Food

Polysorbate 65 is used as a surfactant, often in combination with other emulsifiers for forming oil-in-water emulsions. It is used to hold the fat in ice cream to give dry eating characteristics and to retard the development of fat bloom in chocolate products. It is also used to reduce foam formation during food processing.

Benefits

Polysorbate 65 can be used in combination with mono- and diglycerides of fatty acids (E471) or other polysorbates to provide the optimum balance of emulsion properties. It is soluble in hot and cold water but insoluble in edible oils.

Limitations

Polysorbate 65 is one of the polysorbates included in Annex IV of Directive 95/2/EC as amended by Directives 2003/114/EC and 2006/52/EC, where they are permitted in a range of product categories, with limits in each case. Polysorbate 65 has a waxy, somewhat bitter taste.

Typical Products

Ice cream and frozen desserts, sugar confectionery, cakes and cake mixes, coffee whiteners and whipped creams based on dairy and vegetable fat.

E440	Pectins
	(i) pectin
	(ii) amidated pectin

Sources

Pectins are found in most land plants, especially in fruits and other non-woody tissues. Commercial pectins are currently extracted from fruit solids remaining after juice extraction – in particular, from apple pomace and citrus peel. Other minor sources are sugar beet pulp after the removal of sugar, and sunflower head tissue after removal of the seeds. The choice of source material is limited by availability on a sufficient scale, and by the suitability of the pectins obtained for use in food additive and ingredient functions.

Function in Food

Pectins are used as gelling and thickening agents in a range of mainly acidic foods, most typically fruit products (jams, jellies, industrial fruit preparations for bakery and dairy products, sugar confectionery) but also increasingly in glazes and sauces for savoury products. High-methoxyl pectins are also used as stabilisers of proteins in acidic products such as yoghurts and soya analogues, where heat treatment is required, in ice pops and sorbets, and to improve mouthfeel in drinks (especially low-calorie or low-fruit). Low-methoxyl pectins are also used to gel or thicken desserts, either water- or milk-based.

Benefits

Pectin derived from fruit is the obvious gelling agent to supplement the natural pectin in fruit products. In confectionery, it gives a clear tender gel with good flavour release, which requires no stoving process after depositing. In low-sugar fruit bases, amidated low-methoxyl pectin can give a thixotropic texture, which is pumpable but capable of suspending fruit pieces. Amidated pectin produces completely thermally reversible gels, whilst non-amidated low-methoxyl and high-methoxyl pectins give gels with considerable resistance to melting, and hence bakefast properties. Pectin is an effective stabiliser for acidic protein systems, which does not give excessive viscosity, and is therefore ideal for yoghurt and similar drinks.

Pectins are one form of soluble dietary fibre, and may be used to increase the fibre content of suitable foods and drinks.

Limitations

Both pectin and amidated pectin are generally permitted additives in Annex I of Directive 95/2/EC. Under Directive 2006/52/EC amending Directive 95/2/EC, E440 is not permitted for use in jelly mini-cups.

Typical Products

Jams, marmalades, sugar confectionery (fruit-flavoured and neutral, or thickened), industrial and bakery fillings and toppings, fruit bases for yoghurts, glazes and sauces for ready meals, ice pops and sorbets, yoghurt drinks and soft drinks.

E442   Ammonium phosphatide (Emulsifier YN)

Sources

Ammonium phosphatide is obtained by phosphorylation of a mono- and diglyceride produced from an edible fat. Traditionally, a partially hydrogenated rapeseed oil is used as a fat source.

After the phosphorylation with phosphorus pentoxide, the product is neutralised with ammonia, forming a mixture of ammonium salts of phosphatidic acids.

Function in Food

Ammonium phosphatide is an emulsifier mainly added to chocolate in order to reduce the viscosity of the liquid chocolate, thus making it suitable for further processing such as moulding or enrobing. In chocolate, ammonium phosphatide is found on the surface of the particles, especially on sugar particles, so the friction between the particles is reduced. Ammonium phosphatide also works by dispersing agglomerated particles during the conching process.

Ammonium phosphatide is also added to couverture, ice-cream coatings, and various confectionery products, where it can be used as a substitute for soya lecithin.

Benefits

Ammonium phosphatide has a very neutral flavour profile and does not add any off-flavours to the food products, even when added at high dosages up to 1%. Ammonium phosphatide provides a higher stability against oxidation compared with, for example, soya lecithin.

Its ability to control viscosity makes it possible to reduce the fat content of the final products. In chocolate, ammonium phosphatide works synergistically with the emulsifier PGPR (E476), enabling the manufacturer to obtain an additional saving in the amount of cocoa butter added.

Limitations

Ammonium phosphatide is permitted in Annex IV of Directive 95/2/EC, as amended by Directive 98/72/EC, in cocoa and chocolate products and confectionery made from them.

Typical Products

Ammonium phosphatide is used in chocolate, couverture, ice-cream coatings, confectionery fillings, drinking chocolate and chocolate spreads.

E444   Sucrose acetate isobutyrate (SAIB)

Sources

SAIB is produced by the controlled esterification of sucrose using acetic and isobutyric acid anhydrides. The precise pattern of esterification will depend on the reaction conditions. The molecular weight can vary between 832 and 856. It is a very viscous clear, colourless liquid.

Function in Food

In the preparation of cloudy, flavoured soft drinks, essential oils are often used in an emulsion as part of the flavouring system. SAIB is used to inhibit the coalescence and separation of these oils from the body of the drink. The oils generally have a lower density than the water in the drink and can, unless some preventive action is taken, separate out at the top of the container. SAIB both increases the density of the oil and acts to stabilise the emulsion, usually in conjunction with other water-phase additives, such as gum arabic (E414). The stabilisation is also believed to be aided by charges on the emulsion droplets generated during the emulsification process.

Benefits

SAIB is flavourless and odourless at the levels used in beverages, and is stable to oxidation. It metabolises to sugar, acetic and isobutyric acids.

Limitations

At room temperature, SAIB is a very viscous liquid and must be either warmed to 60 °C or mixed with orange terpenes before use. According to Annex IV of Directive 95/2/EC as amended by Directive 2003/114/EC, SAIB is permitted only in non-alcoholic flavoured cloudy drinks and flavoured cloudy spirit drinks containing less than 15% alcohol by volume up to a maximum of 300 mg/litre.

Typical Products

Cloudy soft drinks.

| E445 | Glyceryl esters of wood rosin |
| | Ester gum |

Sources

Wood rosin is a pale yellow, acidic material extracted from pine wood chips. The major component is abietic acid. The rosin is reacted with glycerol (E422) to give a mixture of di- and triglycerides, which is purified by counter-current steam distillation to yield a hard, clear, pale yellow thermoplastic resin. It is also known as ester gum.

Function in Food

In the preparation of cloudy, flavoured soft drinks, essential oils are often used in an emulsion as part of the flavouring system. E445 is used to modify the properties of the oils so that they remain evenly distributed throughout the drinks during their shelf-life. It is believed to act by increasing the density of the oil, acting as a stabiliser and through charges generated on the droplet surface during the emulsification process.

Benefits

E445 is odourless and tasteless at the concentrations used. It is available as small beads, which allows for improved dispersion when preparing a solution.

Limitations

Under Directive 95/2/EC, as amended by 98/72/EC and 2001/5/EC, glyceryl esters of wood rosin are permitted only in non-alcoholic flavoured cloudy drinks, cloudy spirit drinks in accordance with Council Regulation (EEC) No 1576/89 laying down general rules on the definition, description and presentation of spirit drinks, and cloudy spirit drinks containing less than 15% alcohol by volume to a maximum level of 100 mg/litre, and in the surface treatment of citrus fruit to 50 mg/kg.

Typical Products

Cloudy soft drinks.

E450	Diphosphates
	(i) **disodium diphosphate**
	(ii) **trisodium diphosphate**
	(iii) **tetrasodium diphosphate**
	(iv) **tetrapotassium diphosphate**
	(v) **dicalcium diphosphate**
	(vi) **calcium dihydrogen diphosphate**

Sources

The original source of the diphosphates is phosphate rock, which is mined in areas such as Morocco, Israel, North America and Russia. Yellow phosphorus is extracted from phosphate rock using either a high-energy electrothermal process or an acid extraction. The phosphorus is burnt in an oxygen atmosphere at very high temperatures to produce phosphorus pentoxide. Phosphorus pentoxide is dissolved in dilute phosphoric acid and reacted with sodium, potassium or calcium hydroxide to produce an orthophosphate. The orthophosphates are then combined in a high- temperature condensation reaction to form chains of two phosphate units, (diphosphates).

Sodium, potassium and calcium diphosphates are available. The diphosphates used in food applications are the disodium, trisodium, tetrasodium, tetrapotassium, dicalcium and calcium dihydrogen forms.

Function in Food

The baking industry is the largest user of the diphosphates, where their principal function is that of leavening agent. The acidic diphosphates are used in this application, the most widely used being sodium acid pyrophosphate (disodium diphosphate), which is usually known by its initials SAPP, and the calcium diphosphates.

The phosphates function as stabilisers in meat products, where they work synergistically with salt, interacting with the meat fibres and causing the fibres to expand and retain water within them. They also work with salt to extract the meat proteins, allowing the formation of a meat protein exudate, which will bind meat pieces together in a comminuted or reformed product.

In processed cheese, cheese preparations and cheese-based sauces, the phosphates act as emulsifying salts. In this application, they break the calcium bridges between the cheese protein molecules by means of ion exchange, converting the insoluble cheese protein complexes into individual soluble protein molecules. These protein molecules are then able to emulsify the fat associated with the cheese, in a manner similar to that of sodium caseinate. As this interaction relies on the exchange of sodium or potassium for the calcium associated with the cheese proteins, the calcium phosphates cannot function in this way.

The diphosphates can aid gel formation in products such as instant whips.

Benefits

At least five grades of SAPP are commercially available, differing in their rate of reaction with sodium bicarbonate for use as raising agents. The slower grades are used in large cakes and refrigerated doughs, where consistency of gas release over a long period of time is required, while the faster grades are used in cake doughnuts and small cakes. The use of SAPP increases the alkalinity of the finished cake and increases the rate of browning compared with the use of monocalcium phosphate.

In meat products, the phosphate and salt interaction extracts salt-soluble protein, which binds individual meat pieces together. The meat fibres also expand, allowing greater retention of meat juices, thereby improving succulence.

Without the use of emulsifying salts, such as the phosphates, it is impossible to produce stable processed cheese or cheese-based sauces.

Limitations

Diphosphates are included in Annex IV of Directive 95/2/EC as amended, in which they are permitted in a very wide range of products, with individual limits in each case. The limits are calculated as g/kg P_2O_5.

The diphosphates are fast-acting in meat, but they are the least soluble and application is more difficult. The diphosphates require vigorous action in order to incorporate them into meat systems. They are less suitable for mince-mix systems, where a blend of diphosphates with triphosphates and/or polyphosphates is recommended.

The use of SAPP as a raising agent can result in a distinct aftertaste, which can be minimised by careful adjustment of the acid to bicarbonate ratio.

Calcium phosphates have poor solubility and this limits their application in many food types.

Typical Products

Bakery products, meat products, processed cheese, sauces (especially cheese-based), beverage whiteners, edible ices, icing sugar, dried powdered foods, milk-based drinks (particularly sterilised and UHT drinks) and baking powders.

E451	Triphosphates
	(i) pentasodium triphosphate
	(ii) pentapotassium triphosphate

Sources

The original source of triphosphates is phosphate rock, which is mined in areas such as Morocco, Israel, North America and Russia. Yellow phosphorus is extracted from phosphate rock using either a high-energy electrothermal process or an acid extraction. The phosphorus is burnt in an oxygen atmosphere at very high temperatures to produce phosphorus pentoxide. Phosphorus pentoxide is dissolved in dilute phosphoric acid and reacted with sodium or potassium hydroxide to produce an orthophosphate. The orthophosphates are then combined in a high-temperature condensation reaction to form chains of three phosphate units (triphosphates).

Two triphosphates are available – pentasodium triphosphate (sodium tripolyphosphate) or pentapotassium triphosphate. The pentasodium form is widely available and widely used; the pentapotassium form is less common.

Function in Food

The phosphates function as stabilisers in meat products, where they work synergistically with salt, interacting with the meat fibres and causing the fibres to expand and retain water within them. They also work with salt to extract the meat proteins, allowing the formation of a meat protein exudate, which will bind meat pieces together in a comminuted or reformed product.

In fish and seafood processing, the triphosphates (and polyphosphates) substantially reduce the drip loss on storage, maintaining the succulence of the products and avoiding the dry and fibrous texture otherwise encountered. In contrast to their functionality in meat products, in this application, the phosphates work both with and without salt.

In processed cheese, cheese preparations and cheese-based sauces, the phosphates act as emulsifying salts. In this application, they break the calcium bridges between the cheese protein molecules by means of ion exchange, converting the insoluble cheese protein complexes into individual soluble protein molecules. These protein molecules are then able to emulsify the fat associated with the cheese, in a manner similar to that of sodium caseinate.

Benefits

In meat products, the phosphate and salt interaction extracts salt-soluble protein, which binds individual meat pieces together. The meat fibres also expand, allowing greater retention of water, thereby improving succulence. The triphosphates do not act as quickly as the diphosphates on the meat proteins; they must first be broken down to the diphosphate form by enzymes in the meat. They are, however, more soluble and better suited to dissolution in brine for injection or tumbling of meat, and are far more suitable for mince-mix systems. The optimum results are achieved when they are used as a blend with diphosphates and/or polyphosphates.

In fish and seafood processing, the phosphates help reduce drip loss and dehydration on storage, improving the succulence of the product and avoiding the dry, fibrous nature often associated with these products.

Without the use of emulsifying salts, such as the phosphates, it is impossible to produce stable processed cheese or cheese-based sauces.

As with most phosphate types, the triphosphates are often combined with diphosphates and/or polyphosphates, to give the advantages of solubility and functionality.

Limitations

Diphosphates are included in Annex IV of Directive 95/2/EC as amended in which they are permitted in a very wide range of products with individual limits in each case. The limits are calculated as g/kg P_2O_5.

Typical Products

Meat products, fish and seafood, processed cheese, sauces (especially cheese-based), beverage whiteners, edible ices, icing sugar, dried powdered foods, and milk-based drinks (particularly sterilised and UHT drinks).

E452	Polyphosphates
	(i) sodium polyphosphate
	(ii) potassium polyphosphate
	(iii) sodium calcium polyphosphate
	(iv) calcium polyphosphate

Sources

The original source of polyphosphates is phosphate rock, which is mined in areas such as Morocco, Israel, North America and Russia. Yellow phosphorus is extracted from phosphate rock using either a high-energy electrothermal process or an acid extraction. The phosphorus is burnt in an oxygen atmosphere at very high temperatures to produce phosphorus pentoxide. Phosphorus pentoxide is dissolved in dilute phosphoric acid and reacted with sodium, potassium or calcium hydroxides to produce an orthophosphate. The orthophosphates are then combined in a high- temperature condensation reaction to form chains of two or three phosphate units.

Further polymerisation, to produce longer chain lengths, is achieved by heating in a furnace to form a "glassy" phosphate. This is ground to give a powder, which is composed of a mixture of different chain lengths varying from four units up to thirty or more units. By varying the polymerisation conditions, it is possible to alter the average chain length. These mixtures of phosphates are grouped together under the heading of polyphosphates. There are sodium, potassium and calcium polyphosphates available. The sodium polyphosphates are widely available and widely used; the potassium and calcium polyphosphates are less common.

Function in Food

The phosphates function as stabilisers in meat products, where they work synergistically with salt, interacting with the meat fibres and causing the fibres to expand and retain more water within them. They also work with salt to extract the meat proteins, allowing the formation of a meat protein exudate, which will bind meat pieces together in a comminuted or reformed product.

In fish and seafood processing, the polyphosphates (and triphosphates) substantially reduce the drip loss on storage, maintaining the succulence of the products and avoiding the dry and fibrous texture otherwise encountered. In contrast to their functionality in meat products, in this application, the phosphates work both with and without salt.

In processed cheese, cheese preparations and cheese-based sauces, the phosphates act as emulsifying salts. In this application, they break the calcium bridges between the cheese protein molecules by means of ion exchange, converting the insoluble cheese protein complexes into individual soluble protein molecules. These protein molecules are then able to emulsify the fat associated with the cheese, in a manner similar to that of sodium caseinate. As this interaction relies on the exchange of sodium or potassium for the calcium associated with the cheese proteins, the calcium phosphates cannot function in this way.

Calcium polyphosphates have poor solubility and this may restrict its function in many food types.

Benefits

In meat products, the phosphate and salt interaction extracts salt-soluble protein, which binds individual meat pieces together. The meat fibres also expand, allowing greater retention of water, thereby improving succulence. The solubility of polyphosphates increases with chain length, and polyphosphates are ideally suited to dissolution in brines for injection into meat or for use in mince-mix systems. On the other hand, they do not act as quickly as di- or triphosphates. The longer chain length takes longer to convert to the diphosphate form. Optimum results in all applications are usually achieved with a blend of poly-, di- and triphosphates.

In fish and seafood processing, the phosphates help reduce drip loss and dehydration on storage, improving the succulence of the product and avoiding the dry, fibrous nature often associated with these products.

Without the use of emulsifying salts, such as the phosphates, it is impossible to produce stable processed cheese or cheese-based sauces.

Limitations

Polyphosphates are included in Annex IV of Directive 95/2/EC as amended, in which they are permitted in a very wide range of products, with individual limits in each case. The limits are calculated as g/kg P_2O_5.

Typical Products

Meat products, processed cheese, fish and seafood, sauces (especially cheese-based), beverage whiteners, edible ices, icing sugar, dried powdered foods, and milk-based drinks (particularly sterilised and UHT drinks).

E459	Beta-cyclodextrin

Sources

Beta-cyclodextrin is a cyclic polymer consisting of seven D-glucose units. It is prepared by enzymic modification of starch.

Function in Food

Because of its unique "doughnut" shape, beta-cyclodextrin is able to trap other molecules and protect them against the external environment. Thus it is used to protect sensitive molecules against the effects of heat and light and to reduce losses through evaporation in high-temperature processes. In practice, the material to be protected is mixed with the beta-cyclodextrin in solution and then the mass is dried using a mild process such as a multistage dryer. The dry powder is then used in the product formulation. Typically, the powder will contain 40% encapsulant and 60% beta-cyclodextrin.

Benefits

It is very difficult to add flavours to products made in high-temperature processes because flavour molecules tend to be volatile and are driven off during the process. Beta-cyclodextrin encapsulation can help to overcome this problem and to improve the flavour of products made by processes such as extrusion and cooking particularly by retaining more of the top notes of the flavours. It is also used to protect sensitive flavours such as orange and lime from oxidation during product storage.

Limitations

Beta-cyclodextrin is permitted under Annex IV of Directive 95/2/EC as amended by Directives 98/72/EC and 2003/114/EC for use in foodstuffs in tablet and coated table form to *quantum satis*, and also for encapsulated flavourings in flavoured teas, flavoured powdered instant drinks and flavoured snacks up to specified limits. It is also permitted as a carrier or carrier solvent to a maximum of 1 g/kg of finished foodstuff. Beta-cyclodextrin is expensive and is used only where protection of flavours is important.

Typical Products

Beta-cyclodextrin is used as a flavour carrier in a range of foods, including sugarless confectionery, extruded snacks and frozen prepared meals.

E460	Cellulose
	(i) microcrystalline cellulose
	(ii) powdered cellulose

Sources

Microcrystalline cellulose (MCC) and powdered cellulose are derived from alpha-cellulose, the most abundant natural polysaccharide found in plants and trees. Powdered cellulose is manufactured by bleaching and washing alpha-cellulose before drying and grinding to give fibres 22–110 microns in length.

Microcrystalline cellulose is manufactured by hydrolysing cellulose fibres in acid, leaving crystalline bundles. After bleaching and washing, the cellulose is dried to give aggregates of very porous particles.

Colloidal grades of microcrystalline cellulose are formed by additional wet mechanical attrition to release individual microcrystals. To prevent reaggregation during drying so that the particles may be easily dispersed, microcrystalline cellulose is treated with a water-soluble hydrocolloid such as carboxymethyl cellulose (E466), guar gum (E412), calcium alginate (E404) or xanthan gum (E415). The co-polymer may also modify the end-use properties.

Bacterial cellulose, obtained from the fermentation of *Acetobacter xylinum*, is treated with co-polymers to give a range of products similar to plant-derived microcrystalline cellulose, but this does not have approval for food use in the EU.

Function in Food

Powdered cellulose and powdered microcrystalline cellulose are insoluble particles, which disperse readily in water. Powdered cellulose binds 4–9 times its weight of water and both materials absorb water and oil. These characteristics are the basis of their uses. Both are used to bind water to reduce stickiness and improve the extrusion properties of pasta and puffed snack foods. Powdered cellulose improves the flow properties of pancake batters and retains moisture, reduces fat uptake, and improves gas retention and crumb structure in cakes, muffins and doughnuts. The water- binding properties are used to protect against freeze-thaw damage in surimi and frozen foods. Dispersions of colloidal grades of microcrystalline cellulose are self-suspending above a critical concentration of around 0.25% and form a gel around and above 1%. Aggregates of microcrystalline cellulose and guar gum give body and creaminess to low-fat foods, such as mayonnaise, dressings and milk drinks.

Benefits

Powdered cellulose and powdered microcrystalline cellulose provide opacity and are a source of insoluble fibre in meal replacers and diet foods.

Their ability to bind water and oil allow them to be used as flavour carriers and free-flow aids in instant foods and grated cheese.

Colloidal microcrystalline cellulose is an efficient emulsion and foam stabiliser. The network is stable at all temperatures from chill to boiling point and it improves cling and coating properties. At higher concentrations, the thixotropic gel maintains the shape of extruded foods, prevents ice-crystal growth and freeze-thaw damage during frozen storage, and holds shape when thawed.

Microcrystalline cellulose has the benefit of imparting body and creaminess without gumminess.

Limitations

Powdered cellulose and microcrystalline cellulose disperse with minimal stirring, but colloidal grades must be dispersed using high shear or homogenisation to give a stable network. The dispersion properties are not affected by temperature. In foods with a pH value below 4.5 or more than about 1% salt (sodium chloride), microcrystalline cellulose should be dispersed in water first. A protective colloid, such as xanthan gum, at around 10% of the weight of microcrystalline cellulose, should be added to avoid flocculation and collapse of the stabilising network. Salts or acid should be added last. Dissolved electrolytes or the presence of other water-soluble gums may extend the time for complete dispersion. In milk, it is best to disperse MCC with homogenisation above 100 bar.

Microcrystalline cellulose and powdered cellulose are listed in Annex I of Directive 95/2/EC as amended. All co-polymers of MCC are also included in Annex I of this Directive.

Typical Products

Puffed snack foods, baked goods, instant foods, diet foods, milk drinks, whipping cream, mayonnaise, dressings, frozen desserts and reformed meats.

E461   Methyl cellulose

Sources

Methyl cellulose is manufactured from purified cellulose by reaction with methyl chloride under controlled conditions.

Function in Food

Methyl cellulose is soluble in cold water, where it has thickening properties, but insoluble in hot water, where gels are formed. This allows binding of food products when heated. Films of methyl cellulose exhibit good oil-barrier properties.

Benefits

The hot gelation properties of methyl cellulose are used to reduce boil-out during heating in a range of sauces and fillings. The thermal gelation properties also allow better binding and hence greatly improved shape retention in products such as reformed meats, reformed vegetables, potato products, vegetarian burgers, and dietetic breads. The barrier properties can be used to reduce oil uptake in deep-fried products, both to lower the fat content of the food and to reduce oil losses in processing.

Limitations

Methyl cellulose is a generally permitted additive under Annex I of Directive 95/2/EC.

Typical Products

Soya burgers, sausages, and other formed products, onion rings, potato croquettes, waffles and other formed potato products, gluten-free bakery products, batters, coatings and doughnuts.

E462   Ethyl cellulose

Sources

Ethyl cellulose is the ethyl ether of cellulose, prepared from wood pulp or cotton by treatment with alkali and ethylation of the alkali cellulose with ethyl chloride.

Function in Food

Bulking agent, raising agent.

Benefits

Ethyl cellulose polymers bring binding, film forming and flavour fixative benefits to food products, helping flavours last longer.

Limitations

It is a generally permitted additive under Annex I to Directive 95/2/EC, as amended by Directive 2006/52/EC. It is also listed under Annex V for use as a permitted carrier and carrier solvent.

The acceptable daily intake (ADI) for ethyl cellulose is given by the Joint FAO/WHO Expert Committee on Food Additives as being 'not specified'.

Typical Products

Food supplements, encapsulated flavourings.

E463	Hydroxypropyl cellulose

Sources

Hydroxypropyl cellulose is manufactured by treatment of purified cellulose with propylene oxide under controlled conditions followed by washing to purify the product.

Function in Food

Hydroxypropyl cellulose (HPC) is insoluble in hot water but soluble in cold water. Solutions vary in viscosity depending on the choice of HPC type. It is used as a stabiliser in aerated products. Hydroxypropyl cellulose has good film-forming and barrier properties and is soluble in ethanol.

Benefits

This additive can stabilise whipped toppings at high ambient temperatures. The film-formation properties may be used to give barrier properties, e.g. against oxidation. The ethanol solubility of hydroxypropyl cellulose allows thickening of alcoholic drinks.

Limitations

Hydroxypropyl cellulose is a generally permitted additive under Annex I of Directive 95/2/EC.

Typical Products
   Aerated toppings.

E464	Hydroxypropyl methyl cellulose

Sources
   Hydroxypropyl methyl cellulose (HPMC) is produced from purified cellulose by treatment with methyl chloride and propylene oxide under controlled conditions, followed by washing to purify the product.

Function in Food
   The properties are comparable to those of methyl cellulose, namely it is soluble in cold water to give thickening properties, but insoluble in hot water, where gels are formed, allowing binding of food products when heated. Films of HPMC exhibit good barrier properties.

Benefits
   The hot-gelation properties are used to reduce boil-out of sauces and fillings during heating in a range of sauces and fillings. The thermal gelation properties also improve binding and hence give better shape retention in products such as reformed meats, reformed vegetables, potato products, vegetarian burgers and dietetic breads. The barrier properties can be used to reduce oil uptake in deep-fried products. Hydroxypropyl methyl cellulose has a higher gelation point and viscosity than comparable methyl cellulose types.

Limitations
   Hydroxypropyl methyl cellulose is a generally permitted additive under Annex I of Directive 95/2/EC.

Typical Products
   Soya burgers and sausages and other formed products, onion rings, potato croquettes, waffles and other formed potato products, batters, coatings, doughnuts and gluten-free bakery products.

E465   Methylethyl cellulose

Sources

Methylethyl cellulose (MEC) is produced from cellulose by chemical treatment under controlled conditions, and purified by washing.

Function in Food

This product has surface activity and can stabilise foams in the presence of fat. Thermoreversible gels can also be formed on heating solutions.

Benefits

Solutions of MEC may be whipped to give good overrun. The foams are tolerant to fat and are able to stabilise egg white when fat is present.

Limitations

Methylethyl cellulose is a generally permitted additive under Annex I of Directive 95/2/EC.

Typical Products

Used in non-dairy creams and toppings, aerated desserts and mousses, meringues, mallows and batters.

E466   Carboxymethyl cellulose (Cellulose gum)

Sources

Carboxymethyl cellulose (CMC) is manufactured from purified cellulose by reaction with monochloracetic acid under controlled conditions, followed by washing to purify.

Function in Food

Carboxymethyl cellulose is soluble in hot and cold water and has thickening properties. It also acts as a stabiliser in frozen products. It has protein reactivity, which is utilised to stabilise low-pH dairy and soya products.

Benefits

Carboxymethyl cellulose dissolves in both cold and hot water to give clear, flavourless solutions with a range of viscosity, depending on the choice of CMC grade. The viscosity build-up can be very rapid, especially when fine particle size products are used. This thickening function is used in a range of drinks, sauces and toppings, and in powders to be made up into these products. CMC is also widely used to stabilise fruit pulp in fruit drinks and drink concentrates. Frozen desserts such as ice creams and water ices use CMC to inhibit the growth of ice crystals and maintain a smooth texture. CMC is also used as a water binder, especially in bakery products. It is, in fact, sodium carboxymethyl cellulose, and this ionic character leads to protein reactivity, which is used to stabilise dairy and soya products with pH levels of the order of 4.5.

Limitations

Carboxymethyl cellulose is a generally permitted additive under Annex I of Directive 95/2/EC.

With the finer particle size grades of CMC, the water uptake can be very rapid, leading to clumping. To avoid this, the product should either be preblended with other ingredients such as sugar, or blended with a high-shear mixer.

Typical Products

Soft drinks, dairy drinks, powders and concentrates for drinks, sauces and dressings, ice creams and water ices, bakery products and low-pH dairy products.

E468	Cross-linked sodium carboxymethylcellulose (cross-linked cellulose gum)

Sources

Cross-linked sodium carboxymethylcellulose is the sodium salt of a thermally cross-linked partly O-carboxymethylated cellulose. Cellulose, from wood pulp or cotton fibres, is reacted in sodium hydroxide with sodium monochloroacetate to form sodium carboxymethylcellulose, which, when heated under acid conditions, will cross-link.

Function in Food

Disintegrating agent to accelerate the break-up, dispersion and/or dissolution in water of tablets, capsules or granules.

Benefits

Disintegration gives a faster and better dispersion when a tablet, capsule or granule is added to water. The release of soluble components, such as vitamins, will be accelerated.

Limitations

Cross-linked sodium carboxymethylcellulose is in Annexes IV and V of Directive 95/2/EC, the miscellaneous additives Directive, as amended by Directives 98/72/EC and 2006/52/EC, and may be added to food supplements supplied in solid form up to 30 g/kg, and as a carrier for sweeteners without restriction. Cross-linked sodium carboxymethylcellulose is hygroscopic and should be stored in closed containers in a cool dry place. Efficacy may be slightly reduced by wet granulation processing or by the inclusion of large amounts of other soluble materials.

Typical Products

Sweetener tablets and solid dietary supplements, such as vitamin, fibre and mineral tablets.

E469	Enzymatically hydrolysed carboxy methyl cellulose (Enzymatically hydrolysed cellulose gum)

There is no commercial production of this material.

E470a	Sodium, potassium and calcium salts of fatty acids
E470b	Magnesium salts of fatty acids

Sources

The salts of fatty acids are made by reacting the acids with the appropriate hydroxide. The acids used are principally stearic, palmitic and oleic (see E570). The salts can be used singly or in mixtures.

Function in Food

The salts of fatty acids have a range of uses, usually derived from their fatty acid component. Thus they are free-flow agents, and anticaking and defoaming agents.

Benefits

Magnesium stearate is used to help powders flow during tableting. Other fatty acid salts are used to decrease foam during the processing of beet sugar, as an antitack agent in chewing gum, and as a yeast activity promoter.

Limitations

The salts of fatty acids are generally permitted additives under Annex I of Directive 95/2/EC.

Typical Products

Magnesium stearate is used in tablets.

E471	Mono- and diglycerides of fatty acids

Sources

Mono- and diglycerides occur naturally as food fat constituents, and are also formed from triglycerides, being normal products of fat metabolism, during the digestion and absorption of food. As such, they are always found in conjunction with triglycerides, glycerol and some free fatty acids subject to the manner in which they have been produced.

They are produced commercially by a) heating triglyceride fats with an excess of glycerol, or b) direct esterification of glycerol with fatty acids. The resulting composition is dependent upon the proportion of glycerol and temperature conditions used. The mono-ester is usually in the range 30–60%. The composition of the product will vary according to conditions, but glyceryl monostearate and glyceryl distearate are often major components. In respect of their listing under miscellaneous additives status, the content of mono- and diesters must not be less than 70%. This provides the opportunity for a wide range of compositional types, where specific applications may be required. Accordingly, these products will vary in their appearance from pale straw to brown oily liquids, to white or slightly off-white hard waxy solids. The solids may be in the form of flakes, powders or small beads. Typically, these products are insoluble in water but can form stable hydrated dispersions.

Function in Food

Emulsifiers are used to disperse fat droplets in water or water droplets in fat. Because they act at the surface between the fat and the water, they are also known as surface-active agents or surfactants. Monoglycerides and mixtures of mono- and diglycerides are by far the most important commercially of all the food surfactants known; in Europe, they represent no less than 50% of the total food emulsifier market and, in addition, the monoglycerides are important intermediates in the manufacture of DATEMs (diacetyl tartaric acid esters of monoglycerides) and other emulsifiers.

Mono- and diglycerides are used widely in a great many products and are the surfactant type most used in bread.

Performance characteristics are controlled by the skilful combining of alpha-mono-, beta-mono-, di- and tri-glyceryl esters of mixtures of fatty acids.

In bread, the effectiveness of the emulsifier is dependent upon its total monoester content as, in this application, the performance of the alpha- and beta-fractions are similar and superior to either the di- or triglycerides.

Benefits

In the production of bread, the contribution made by the surfactant is to enable the gluten in the dough to remain plastic and pliable so that, during the kneading process, the strands of gluten can form a smooth extensible film, ensuring that the correct texture is produced in the finished product.

As a general rule, volume and texture are of the utmost importance in baking. For example, it can be shown that, in cake making, the air bubbles in the batter that contribute to volume are enclosed in films of protein in which the fat is dispersed. The action of the surfactant is to improve the production of the initial air bubbles, ensuring their uniformity and thereby an improved texture of the finished baked product.

Limitations

The mono- and diglycerides are generally permitted additives in Annex I of Directive 95/2/EC.

Typical Products

Bread, cakes and other baked goods; cereals, puddings; fresh pasta, instant (mashed) potatoes; frozen desserts, ice cream and soft-serve; confectionery, e.g. chewing gums, toffees, caramels; and fats, e.g. margarines and shortenings.

E472a	Acetic acid esters of mono- and diglycerides of fatty acids (acetems)

Sources

Acetems are made by reacting mono- and diglycerides of fatty acids with acetic acid in various proportions. The properties of the acetems are decided by selection of the fatty acids used to make the glyceride backbone, together with control of the number of hydroxyl groups remaining – characterised as the proportions of mono- and diglycerides. The extent to which the remaining hydroxyl groups of the fatty acid glycerides are reacted with acetic acid further modifies both the melting point and the hydrophile–lipophile balance (HLB). The HLB of acetems is usually quite low, around 2 or 3, indicating that the affinity for oils or fats is much greater than it is for water.

The acetic acid used is from fermented or synthetic sources and the fatty acids and glycerol can be of animal or vegetable origin.

Acetems are available in liquid, pasty and solid forms, with a wide range of melting points.

Function in Food

Acetems are used as emulsifiers, stabilisers and solubilisers, to modify mouthfeel and texture, for protective coatings and to modify the plasticity of fats.

Benefits

In blends for whipped toppings or as aerating/emulsifying agents for cakes and sponges, acetems provide emulsification and stabilisation for the aqueous foams of protein, fat and sugar. In chewing confectionery, they are used to adjust juiciness, texture and stickiness. Acetems that are solid at room temperature are used to coat and protect food such as sausages, fruit and cheese. Apart from preventing microbiological contamination, the barrier controls moisture migration and provides a removable surface for labelling. Films of appropriate acetems can reduce uptake of taints and extend the shelf-life of products such as liver sausage by preventing contact with the air, which can lead to harmless but ugly surface discoloration. These films can also retain protective atmospheres. In the preparation of jams and marmalades, acetems are used as an antifoam to aid filling and present a neat, unbubbled surface. Acetems are also used to manipulate the melting point and plasticity of fats.

Limitations

Acetems are generally permitted additives in Annex I of Directive 95/2/EC.

Typical Products
Cakes and sausages.

E472b	Lactic acid esters of mono- and diglycerides of fatty acids (lactems)

Sources
Lactems are made by reacting mono- and diglycerides of fatty acids with lactic acid in various proportions. The properties of the lactems are decided by selection of the fatty acids used to make the glyceride backbone, together with control of the number of hydroxyl groups remaining – characterised as the proportions of mono- and diglycerides. The extent to which the remaining hydroxyl groups of the fatty acid glycerides are reacted with lactic acid further modifies both the melting point and the hydrophile–lipophile balance (HLB). The HLB of lactems is higher than that of acetems, around 3 or 4, indicating a slightly greater affinity for water.

The lactic acid used is from fermented or synthetic sources and the fatty acids and glycerol can be of animal or vegetable origin.

Lactems are available in liquid, pasty and solid forms.

Function in Food
Lactems are used as emulsifiers in oil-in-water emulsions, frequently in combinations with more hydrophilic emulsifiers to produce blends capable of making stable water-in-oil emulsions. They are used to improve the incorporation and distribution of air in whipped systems such as cakes and mousses.

Benefits
Lactems are used in combination with other emulsifiers in whipped topping concentrates, and as aerating and emulsifying agents for cakes and sponges to produce a narrow pore size distribution in the crumb. In mousses these combinations of emulsifiers are used to produce consistent aeration and to maximise volume. Lactems are also used in baking margarines.

Limitations
Lactems are generally permitted additives in Annex I of Directive 95/2/EC.

Typical Products
Cakes and whipped toppings.

E472c	Citric acid esters of mono- and diglycerides of fatty acids (citrems)

Sources

Citrems are made by reacting mono- and diglycerides of fatty acids with citric acid in various proportions. The properties of the citrems are decided by selection of the fatty acids used to make the glyceride backbone, together with control of the number of hydroxyl groups remaining, usually expressed as the proportions of mono- and diglycerides. The extent to which the remaining hydroxyl groups of the fatty acid glycerides are reacted with citric acid further modifies both the melting point, solubility and the hydrophile–lipophile balance (HLB). The HLB of citrems is around 6–10, higher than that of lactems at 3 or 4 and acetems at 2 to 3.

The citric acid used is produced by fermentation and the fatty acids and glycerol can be of animal or vegetable origin.

Function in Food

Citrems are used as emulsifiers to prevent separation of fat during cutting or chopping and to stabilise emulsions in cooked products such as liver sausage. They are used to reduce spattering of margarines during frying. They are also used in combination with other emulsifiers in sauces and dressings and as solubiliser and synergist for antioxidants. Special citrems are used in the production of dried yeast to protect the yeast cells during drying.

Benefits

Because they contain free acid groups, their HLB value is affected by pH, rising to a maximum around pH 7.

Limitations

Citrems are generally permitted additives in Annex I of Directive 95/2/EC as amended.

Typical Products

Sausages and frying margarines, cocoa and chocolate products.

| E472d | Tartaric acid esters of mono- and diglycerides of fatty acids (tatems) |

Sources

Tatems are formed by the reaction of mono- and diglycerides of fatty acids with tartaric acid. The fatty acids and glycerol can be of animal or vegetable origin.

Owing to the high content of expensive tartaric acid, tatem is certainly the most costly of the E472 series and, because better and more convenient functionality is available from datems or others in this group of emulsifiers, it is rarely produced commercially.

Tatems vary in physical form from sticky liquids to solids and can be white to pale yellow in colour.

Function in Food

Tatem is an emulsifier and stabiliser.

Limitations

E472d is a generally permitted additive in Annex I of Directive 95/2/EC.

| E472e | Mono- and diacetyl tartaric acid esters of mono- and diglycerides of fatty acids (datems) |

Sources

Datems are made by reacting mono- and diglycerides of fatty acids with tartaric acid in various proportions and then adding acetic anhydride to acetylate the free hydroxyl groups of the tartaric acid. The properties of the datems are decided by selection of the fatty acids used to make the glyceride backbone and the control of the subsequent steps of the reaction.

The tartaric acid used is a by-product of the wine industry; acetic acid comes from fermented or synthetic sources and the fatty acids and glycerol can be of animal or vegetable origin.

Datems are available in liquid, pasty and solid forms, with a wide range of melting points.

Variation in the balance of supply and demand for the natural tartaric acid is the principal cause of the price volatility of commercial datems.

Function in Food

Datems are used as emulsifiers in a wide range of food products, but particularly where there is potential for interaction with protein, such as in wheat-based baked goods and egg-containing emulsions such as mayonnaise. The formation of hydrogen bonds between the datem and the gluten proteins in wheat flour strengthens the gluten network. Datems also stabilise egg proteins and render them less susceptible to coagulation under conditions of heat or shear.

Benefits

In the major use area of yeast-raised baked goods, datems work as classical emulsifiers of oil and water to ensure even and stable distribution of lipids, but they also improve dough performance in a number of ways. Mixing tolerance, that is to say the length of time for which the dough can be mixed or manipulated without over-extension and loss of condition, is increased by the use of datems. Fermentation tolerance is also improved as the period of time for which a dough remains at or near peak of volume development is lengthened. The volume of baked goods can also be increased because of better gas retention as doughs are fermented and handled.

Limitations

Datems are generally permitted additives in Annex I of Directive 95/2/EC.

Typical Products

Baked goods.

E472f	Mixed acetic and tartaric acid esters of mono- and diglycerides of fatty acids (matems)

Sources

Matems are made by reacting mono- and diglycerides of fatty acids with a mixture of tartaric acid and acetic acid in various proportions. Since acetic acid can also react with tartaric acid to form acetylated tartaric acid, there are effectively three acids in this reaction mixture competing for the free hydroxy groups of the fatty acid glyceride. A considerable number of different reaction products is thus possible, and very tight control of the processing conditions is required to make a product of consistent quality.

The HLB values of such complex mixtures are dependent on the ratios of their components and, since the behaviour of the products is also affected by the extent to which free acid groups are neutralised with inorganic bases, the HLB value must be determined for each overall specification.

The acetic acid comes from fermented or synthetic sources and the fatty acids and glycerol can be of animal or vegetable origin.

Matems vary in physical form from sticky liquids to solids and may be white to pale yellow in colour.

Function in Food

Since they can exhibit some of the characteristics of both acetems and datems, their potential applications for emulsification and stabilisation are numerous. However, they are not widely manufactured or used.

Benefits

In the principal area of use, baked goods, matems work as classical emulsifiers of oil and water but can also interact with gluten proteins. They improve the tolerance of doughs to extended mixing and increase the period of time for which a dough remains at or near the peak of volume development. Matems are also used in making rusks and for ready-to-mix flours.

Limitations

E472f is a generally permitted additive in Annex I of Directive 95/2/EC.

Typical Products

Baked goods.

E473	Sucrose esters of fatty acids

Sources

Sucrose esters of fatty acids are prepared by esterifying one or more of the (primary) hydroxyl groups of sucrose with the methyl and ethyl esters of food fatty acids or by extraction from sucroglycerides (E474). Depending upon the degree of esterification, a wide range of sucrose esters is obtained, covering the major part of the hydrophile–lipophile balance (HLB) scale.

Function in Food

Sucrose esters are generally used as emulsifiers in (mainly) oil-in-water emulsions, but can also be used as texturisers in, e.g. fine bakery wares or to stabilise foam in dairy products and their analogues. Furthermore, sucrose esters of fatty acids exhibit some selective antimicrobial activity, especially against Gram-positive bacteria.

Benefits

Sucrose esters of fatty acids are off-white, free-flowing powders. They are neutral in taste and odour, and do not influence the taste of other ingredients present in a formulation. Being heat-stable, heating to temperatures up to 185 °C is possible without any negative effect on performance. While being vegetable-derived, sucrose esters of fatty acids are of constant and high quality.

Because they are efficient surfactants, sucrose esters in general can be used at low dosage levels.

Limitations

Sucrose esters are permitted in Annex IV of Directive 95/2/EC in a large variety of products, which is further extended under Directive 98/72/EC, as amended by Directive 2006/52/EC.

Sucrose fatty acid esters are stable at temperatures up to 185 °C and at pH levels between 4 and 8. In addition, there are no known incompatibilities of sucrose fatty acid esters with other food ingredients.

Typical Products

Ice cream, fine bakery wares, cream analogues, canned coffee, beverage whiteners, non-alcoholic almond and coconut drinks, powders for hot beverages, sauces and confectionery.

E474	Sucroglycerides

Sources

Sucroglycerides are emulsifiers derived from natural ingredients (transesterification of natural triglycerides by sucrose). Different sources of triglycerides can be used, such as palm oil, rapeseed oil, coconut oil, castor oil, hydrogenated palm oil or tall oil.

Function in Food

Sucroglycerides are good emulsifiers and dispersing aids. They help to control crystallisation and improve texture. They have good synergy with proteins.

Benefits

These speciality emulsifiers are less expensive than sucrose esters. The different types of triglyceride allow development of a wide range of products with different customised functionalities.

Limitations

Sucroglycerides are permitted in Annex IV of Directive 95/2/EC as amended by Directives 98/72/EC and 2006/52/EC in a large variety of products with individual limits.

Typical Products

Non-alcoholic drinks, bakery products and ice creams.

E475   Polyglycerol esters of fatty acids

Sources

Polyglycerol esters of fatty acids are made by reaction of polyglycerol and the fatty acids, either alone or in mixtures. Polyglycerol itself is derived from glycerol.

Function in Food

Polyglycerol esters of fatty acids are water-dispersible and more polar than monoglycerides, so they are used to hold water in fat emulsions. They are also used to facilitate aeration in cake mixes.

Benefits

The esters act synergistically with other emulsifiers to reduce the amount of emulsifier required in products such as desserts and to reduce the amount of hard fat needed in margarine blends. They also reduce the tendency of low-fat spreads to weep on storage.

Limitations

Polyglycerol esters of fatty acids are permitted in Annex IV of Directive 95/2/EC, as amended by Directive 2006/52/EC for a number of foodstuffs, including bakery wares, emulsified products and chewing gum, with individual limits in each case.

Typical Products

The polyglycerol esters are used in cakes and gateaux, and frozen cheesecake.

E476	Polyglycerol polyricinoleate (PGPR)

Sources

PGPR is a mixture of partial esters of polyglycerol with linearly interesterified castor oil fatty acids (ricinoleic acid). The polyglycerol moiety is predominantly di-, tri- and tetra-glycerol.

The glycerol is typically derived from vegetable oil. The ricinoleic acid is derived from vegetable castor oil.

Function in Food

PGPR is used to modify the flow characteristics of chocolate and, because it is an efficient surfactant, to stabilise water-in-oil emulsions such as low-fat spreads.

Benefits

PGPR acts as a viscosity modifier in chocolate- and cocoa-based products. Rheology in chocolate is complex, and the flow behaviour is described by two parameters – yield value and plastic viscosity. Yield value relates to the force needed to start liquid chocolate moving and plastic viscosity to the force necessary to keep it moving.

PGPR decreases yield value and therefore improves the flow properties and handling of chocolate for coating, moulding and block chocolate production. It also decreases the risks of defects such as air bubbles in the finished product.

PGPR also has a synergistic effect with lecithin, which has a beneficial influence on plastic viscosity. Use of PGPR allows the reduction of fat levels in the product.

Limitations

PGPR is permitted in cocoa-based confectionery and low-fat spreads under Annex IV of Directive 95/2/EC as amended by Directive 98/72/EC.

Typical Products

Chocolate and chocolate products, and low-fat spreads.

E477   Propane 1,2 diol esters of fatty acids

Sources
Propane 1,2 diol esters of fatty acids are made by reaction of the acids with 1,2 epoxypropane or by reaction of propylene glycol (propane 1,2 diol) and oils such as soya-bean oil. Typically, the fatty acids are palmitic and stearic and the process produces a mixture of mono- and diesters, which can be distilled to produce up to 90% monoester.

Function in Food
The esters are emulsifiers used to improve whippability of powdered desserts and the texture and volume of cakes.

Benefits
The propane 1,2,diol esters of fatty acids are synergistic with other emulsifiers, such as mono- and diglycerides of fatty acids (E471).

Limitations
The propane-1,2-diol esters of fatty acids are permitted in a limited range of products, as specified in Annex IV of Directive 95/2/EC as amended by Directive 2006/52/EC.

E479b   Thermally oxidised soya-bean oil interacted with mono- and diglycerides of fatty acids

Sources
As the name suggests, this product is made by first heating soya oil in the presence of air to initiate polymerisation. The resulting complex mixture is then reacted with a mono- and diglyceride mixture, such as E471.

Function in Food
The mixture is used to stabilise fat emulsions used for frying on a hot plate or griddle.

Benefits
E479b provides a stable emulsion that has low viscosity but coats the frying surface, does not char during frying, and gives good release of the product from the griddle, leaving it clean.

Limitations

E479b is permitted in Annex IV of Directive 95/2/EC, but only for use in fat emulsions for frying.

E481	Sodium stearoyl-2-lactylate
E482	Calcium stearoyl-2-lactylate

Sources

The stearoyl lactylate is made by reacting together stearic and lactic acids and neutralising the resulting acid with the appropriate base.

Function in Food

The stearoyl lactylates are hydrophilic emulsifiers, which are used to disperse fat evenly in water-based formulations.

Benefits

They are used to distribute fat in bread dough to give a uniform crumb structure and improve keeping quality. They are also used in emulsifier blends to improve fat suspension during spray drying of fat powders. They are ionic and bind to both proteins and starches.

The calcium salt in particular is used to increase strength and volume in bread and to increase the tolerance to processing.

Limitations

The stearoyl lactylates are permitted in a range of foods under Annex IV of Directive 95/2/EC as amended by Directive 2006/52/EC.

Typical Products

Beverage whiteners, bread and low-fat spreads.

E483	Stearyl tartrate

Sources

Stearyl tartrate is made by reacting stearyl alcohol with tartaric acid (E334). The alcohol is either synthesised from ethylene or prepared from stearic acid.

Function in Food

Stearyl tartrate is used as an emulsifier.

Limitations

Under Directive 95/2/EC, stearyl tartrate is permitted only in bakery wares to a maximum of 4 g/kg and desserts to a maximum of 5 g/kg.

E491	**Sorbitan monostearate**
E492	**Sorbitan tristearate**
E493	**Sorbitan monolaurate**
E494	**Sorbitan monooleate**
E495	**Sorbitan monopalmitate**

Sources

The sorbitan esters are produced by the reaction of the appropriate fatty acid with hexitol anhydride, which is itself derived from sorbitol.

Function in Food

The sorbitan esters are non-ionic emulsifiers, which have a range of hydrophile–lipophile balance (HLB) values from 2 to 8. The HLB system indicates whether emulsifiers will tend to favour oil-in-water or water-in-oil emulsions, and a figure lower than 6 indicates a preference towards water-in-oil. The tristearate is particularly lipophilic.

Benefits

The sorbitan esters are used to hold aqueous solutions in suspension in fatty materials. Thus they are used to disperse aqueous additives in ice cream, fat spreads and desserts, to provide stable emulsions of fat for spray drying, as beverage whiteners or fat powders, and to inhibit staling in bakery products by interrupting the structure. They are also used as antifoaming agents in the production of beet sugar, boiled sweets and preserves, and to modify fat crystal structure in chocolate to inhibit the development of the storage defect known as "bloom".

The esters tend to be used in mixtures, with each other or with the polyoxyethylene sorbitan esters (E432 to 435) to generate the optimum HLB value for the particular food system.

Limitations

The sorbitan esters are listed in Annex IV of Directive 95/2/EC, as amended by Directive 2006/52/EC, permitted in a range of products each with their individual maximum concentration.

Typical Products

The sorbitan esters are used in cake mixes and fat spreads.

E500	Sodium carbonates
	(i) sodium carbonate
	(ii) sodium hydrogen carbonate (sodium bicarbonate)
	(iii) sodium sesquicarbonate

Sources

Sodium bicarbonate is made industrially from brine and limestone using the ammonia soda process. It is purified by repeated crystallisation.

Sodium carbonate is made by heating the impure sodium bicarbonate. It is also produced in the USA from sodium sesquicarbonate ore by heating followed by leaching with warm water.

Sodium sesquicarbonate is mined in the USA, where it is known as "trona". It is also produced by crystallising a mixture of sodium carbonate and sodium bicarbonate.

Function in Food

Of the carbonates, the most common is sodium bicarbonate. It is also known as bicarbonate of soda or baking soda. It is used in baking powder, to generate carbon dioxide by mixing it with an acidic material such as tartaric acid (E334) or one of the acidic phosphates. Sodium bicarbonate does decompose thermally and can be used alone as a raising agent.

Sodium carbonate is also used as a raising agent in cakes, in combination with, for example, sodium aluminium phosphate (E541).

The sodium carbonates are also used to modify the acidity of products and to stop the hydrolysis reaction in the production of invert sugar.

Benefits

The sodium carbonates are soluble in cold water, readily available and inexpensive. Their rate of reaction with acids can be varied by changing the particle size, and both carbonate and bicarbonate are available in a number of granular sizes. They are also available as granules coated (encapsulated) with fat or magnesium stearate for applications where the reaction needs to be inhibited until later in the process.

The bicarbonate is also used alone, generating carbon dioxide by the action of heat at temperatures as low as 60 °C.

Limitations

When used in excess, the bicarbonate can leave a soapy taste in the product. The particle size of the carbonate has to be chosen with care since use of too large a particle can result in there being unreacted carbonate in the final product.

The carbonates are generally permitted additives under Annex I of Directive 95/2/EC.

Typical Products

The carbonates are used in a wide range of baked goods including pastries, cakes, waffles, cookies and scones.

E501	Potassium carbonates
	(i) potassium carbonate
	(ii) potassium hydrogen carbonate (potassium bicarbonate)

Sources

Both potassium carbonate and bicarbonate are prepared by passing carbon dioxide into potassium hydroxide.

Function in Food

Potassium carbonate is used as a raising agent in conjunction with an acidic material such as sodium aluminium phosphate (E541). It is also used in the alkalisation of cocoa powder.

Potassium bicarbonate can also be used in baking powder to generate carbon dioxide by mixing it with an acidic material such as tartaric acid (E334) or disodium diphosphate (E450(i)).

The tricarbonate is also used alone, generating carbon dioxide by the action of heat at temperatures as low as 60 °C.

Benefits

The potassium carbonates are used as raising agents, where it is necessary to restrict the amount of sodium or enhance the potassium in the product. Potassium carbonate is more soluble than sodium carbonate.

In the alkalisation of cocoa powder, the powder is reacted with a base to deepen the colour and increase the intensity of flavour. A number of bases are used for this purpose, each having its particular advantages. Potassium carbonate is considered to give a better colour than sodium carbonate.

Limitations

Potassium carbonate releases carbon dioxide only when used in conjunction with an acid. It is thus less convenient to use than sodium carbonate.

When used in excess, potassium bicarbonate can leave a soapy taste in the product.

Potassium bicarbonate is more expensive and requires higher usage rates than sodium bicarbonate.

The carbonates are generally permitted additives under Annex I of Directive 95/2/EC.

Typical Products

Low-sodium crackers or biscuits and energy bars. Potassium carbonate is also used in the treatment of hops, in cocoa powder used for baking or chocolate drinks, and in gingerbread.

E503	Ammonium carbonates
	(i) ammonium carbonate
	(ii) ammonium hydrogen carbonate (ammonium bicarbonate)

Sources

Ammonium bicarbonate is prepared by passing carbon dioxide into ammonia solution. The bicarbonate precipitates and is washed and dried. The carbonate is prepared by subliming a mixture of ammonium sulphate and calcium carbonate together and is actually a mixture of the bicarbonate and the carbamate.

Function in Food

The ammonium carbonates are frequently used as mixtures as they have similar performance. Both are used in baking to generate carbon dioxide by the action of heat or with acids. They can also be used in the alkalising of cocoa powder.

Benefits

The ammonium carbonates are particularly useful because they break up on heating to only 60 °C, generating both carbon dioxide and ammonia, and leaving no residue in the product. They are both readily soluble in water.

Limitations

Ammonium carbonates tend to be used only in thin products with a final moisture content below 5%. Products outside these constraints, such as cakes or soft cookies, can retain ammonia within the final product with an adverse impact on quality.

Ammonium carbonates are generally permitted additives under Directive 95/2/EC. The carbonate must be stored in sealed containers as it loses ammonia and carbon dioxide on exposure to the air, to leave the bicarbonate. Both products must be stored at or below room temperature.

Typical Products

The ammonium carbonates are used as raising agents in the manufacture of biscuits and crackers, and in sugar confectionery.

E504	Magnesium carbonates
	(i) magnesium carbonate
	(ii) magnesium hydrogen carbonate (magnesium bicarbonate)

Sources

Magnesium carbonate is made from dolomite, a naturally occurring mineral.

Magnesium bicarbonate is made by passing carbon dioxide into a magnesium hydroxide slurry at high pressure.

Function in Food

Magnesium carbonate is used as a source of carbon dioxide, using either heat or acid. It is also a source of magnesium in fortified products and a free-flow agent in table salt.

It has been used as a pharmaceutical antacid and can be used to reduce the acidity of foodstuffs.

Benefits

Magnesium carbonate is inexpensive, is available as a fine powder and is not hygroscopic, all of which make it useful as a free-flow agent.

Limitations

Magnesium carbonate is a generally permitted additive under Directive 95/2/EC.

Typical Products

Magnesium carbonate is used in cheese, ice cream and table salt.

E507　　Hydrochloric acid

Sources

Hydrochloric acid is made industrially from salt.

Function in Food

Hydrochloric acid is a strong acid and is used to increase the acidity of formulations and in the hydrolysis of large molecules such as proteins.

Benefits

Hydrochloric acid is a strong acid and is very cost-effective. It is often used because it has a less acid taste than other acids, such as citric.

It is also used in the production of invert sugar from sucrose and in the production of glucose syrups from starch.

It is used in the hydrolysis of vegetable proteins and has the advantage that the usual product of neutralisation is salt, which enhances the taste of savoury products such as the hydrolysed proteins.

Limitations

Concentrated hydrochloric acid is corrosive and it must be handled with great care to avoid contact with skin. Once it is diluted in the food, it is harmless. Because it is a strong acid it usually needs to be mixed in rapidly to avoid local decreases in pH, which could have irreversible effects.

The degree of hydrolysis of starch that can be achieved with hydrochloric acid is limited, and enzymes have to be used to produce high-sweetness syrups.

Hydrochloric acid is generally permitted under Directive 95/2/EC.

Typical Products

Invert (golden) syrup, glucose syrup and hydrolysed vegetable proteins.

E508   Potassium chloride

Sources
Potassium chloride is purified from a natural mineral source.

Function in Food
Potassium chloride has a taste similar to salt and is used to provide saltiness in products where low sodium content is required. It is also used on its own, and in combination with other ingredients, in table-top salt replacers.

Benefits
Potassium chloride is used in products for people who wish to limit their sodium intake as it contains no sodium ions.

Limitations
Potassium chloride does not taste the same as common salt (sodium chloride) and is not a complete replacement.

Potassium chloride is a generally permitted additive under Directive 95/2/EC.

Typical Products
Potassium chloride is used in table-top salt replacers and in dietetic foods.

E509   Calcium chloride

Sources
Calcium chloride is extracted from natural brines or manufactured as a by-product of the production of sodium carbonate.

Function in Food
Calcium chloride is used as an aqueous solution to provide a source of calcium ions.

Benefits
Calcium chloride is soluble in water, which makes it a good source of calcium ions in solution. These ions are used for a number of purposes depending on the product. In brewing, they modify the hardness of the water; in canned

vegetables and vegetable products, they improve texture by reacting with the natural pectin, and they can also be used to cross-link alginate gels.

Calcium chloride is also used to aid coagulation in cheese manufacture and in the extraction of alginates from seaweed.

Limitations

Calcium chloride is a generally permitted additive under Directive 95/2/EC as amended.

Typical Products

Canned and bottled fruit and vegetables such as carrots, kidney beans, gherkins, olives and pickles and ketchup, and in cheese.

E511   Magnesium chloride

Sources

Magnesium chloride is available as a mineral ore, from underground brines, and is made from the natural mineral dolomite by reaction with hydrochloric acid.

Function in Food

Magnesium chloride is used in the preparation of water for brewing and as a source of magnesium in fortified products.

Benefits

Magnesium chloride is inexpensive and readily soluble in water.

Limitations

Magnesium chloride is a generally permitted additive under Directive 95/2/EC.

Typical Products

None known.

E512   Stannous chloride

Sources

Stannous chloride is made by reacting tin with either chlorine or hydrochloric acid under the appropriate conditions.

Function in Food

Stannous chloride is used to maintain the colour of processed asparagus.

Limitations

Under Directive 95/2/EC, stannous chloride is permitted only in canned and bottled white asparagus to a level of 25 mg/kg.

Stannous chloride absorbs oxygen from the air and should be kept in a tightly sealed container in a cool place.

E513   Sulphuric acid

Sources

Sulphuric acid is made industrially from sulphur dioxide.

Function in Food

Sulphuric acid is a strong acid and is used to increase the acidity of formulations. It is also used in the production of invert sugar.

Benefits

Sulphuric acid is a strong acid and is very cost-effective. The salts formed on neutralising it with common bases have little flavour.

Limitations

Concentrated sulphuric acid is corrosive and it must be handled with great care to avoid contact with the skin. Once it is diluted in the food, it is harmless. Because it is a strong acid, it usually needs to be mixed in rapidly to avoid local decreases in pH, which could have irreversible effects.

Sulphuric acid is a generally permitted additive under Directive 95/2/EC.

Typical Products

None known.

E514	Sodium sulphates
	(i) sodium sulphate
	(ii) sodium hydrogen sulphate (sodium bisulphate)

Sources

Sodium sulphate is produced as a by-product of a number of processes using sulphuric acid. It is also mined as Glauber's salt and purified by recrystallisation.

The bisulphate is made by further reaction of the sulphate with sulphuric acid.

Function in Food

Sodium sulphate is used in colours to standardise the colour strength of the powder.

Sodium bisulphate is also used as an acid in raising agents.

Benefits

Colours, whether natural, nature-identical or synthetic, do not have exactly the same intensity in every batch. Sodium sulphate is used as a neutral material to be blended with the colour to ensure that batches as sold are of consistent intensity.

Limitations

The sodium sulphates are generally permitted additives under Annex I of Directive 95/2/EC.

Typical Products

None known.

E515	Potassium sulphates
	(i) potassium sulphate
	(ii) potassium hydrogen sulphate (potassium bisulphate)

Sources

Potassium sulphates are made by partial or complete neutralisation of sulphuric acid by potassium hydroxide, or by reaction of the acid with potassium chloride.

Function in Food

Potassium bisulphate is used as an acidic material in raising agents. It is used as a replacement for sodium sulphate in products where it is required to reduce the sodium level.

Limitations

The potassium sulphates are generally permitted additives in Annex I of Directive 95/2/EC.

Typical Products

None known.

E516	Calcium sulphate

Sources

Calcium sulphate is a naturally occurring mineral, also known as Plaster of Paris. It is also a by-product of a number of manufacturing processes. Both anhydrous and hydrated forms are available.

Function in Food

Calcium sulphate is used to provide a source of calcium ions.

Benefits

Calcium sulphate is used in the preparation of water for brewing to provide both calcium and sulphate ions, which are present in naturally hard water. In canned fruit and vegetables, it is also used to provide calcium ions for reaction with natural cell-wall pectin to maintain the firmness of the pieces. In baking, it helps bubble stability and cell strength.

Limitations

Calcium sulphate is barely soluble in water.

Calcium sulphate is a generally permitted additive under Directive 95/2/EC.

Typical Products

Wafer biscuits, bread, beer, canned fruit and vegetables and tableted products.

E517	Ammonium sulphate

Sources

Ammonium sulphate is produced by passing ammonia gas into sulphuric acid solution.

Limitations

Ammonium sulphate is permitted only as a carrier according to Annex V of Directive 95/2/EC.

E520	Aluminium sulphate
E521	Aluminium sodium sulphate
E522	Aluminium potassium sulphate
E523	Aluminium ammonium sulphate

Sources

Aluminium sulphate (alum) is manufactured by the reaction of sulphuric acid with naturally occurring aluminium oxide or as a by-product in the manufacture of alcohols.

The mixed sulphates are prepared by mixing concentrated solutions of the two components and allowing them to crystallise as they cool.

Function in Food

Aluminium sulphate is used to improve the resistance of the conalbumin fraction of egg white to denaturation during heat treatment, and thus to preserve the whipping properties of dried egg white. It is not in common use.

Limitations

Under Annex IV of Directive 95/2/EC, aluminium sulphate is permitted (alone or in combination with mixed sulphates E521-E523) only in crystallised fruit and vegetables to a maximum of 200 mg/kg and in egg white to a maximum of 30 mg/kg.

The SCF has agreed an ADI for aluminium from all sources, including additives, of 7 mg/kg body weight.

E524	Sodium hydroxide

Sources

Sodium hydroxide is manufactured industrially by the electrolysis of salt.

Function in Food

Sodium hydroxide is strongly alkaline and is used to decrease the acidity (raise the pH) of food formulations.

Benefits

Because it is a strong base, sodium hydroxide is very cost-effective and is used at very low levels. It is used to neutralise acid and to stop the reaction in the production of invert sugar. It is also used in the alkalisation of cocoa powder and in the hydrolysis of proteins.

It is used in potato processing to improve the efficiency of peeling.

Limitations

Sodium hydroxide is available as both a solid (pellets or flakes) and a concentrated liquid, both of which are very caustic, and precautions need to be taken to avoid these materials coming into contact with the skin. Once diluted in the food it is harmless. Because it is a strong base, great care needs to be taken in its use to avoid severe local increases in pH, which could have irreversible effects.

Sodium hydroxide should be kept in sealed containers because it absorbs water and carbon dioxide from the atmosphere.

Sodium hydroxide is a generally permitted additive under Directive 95/2/EC.

Typical Products

Jams, milk drinks and cocoa powder used for baking or chocolate drinks.

E525	Potassium hydroxide

Sources

Potassium hydroxide is manufactured industrially by electrolysis of naturally occurring potassium chloride.

Function in Food

Potassium hydroxide is a strong base and is used to reduce acidity (increase pH) in foods.

Benefits

Potassium hydroxide is more expensive than sodium hydroxide but is used instead of sodium hydroxide in formulations where it is important to limit the amount of sodium in the final product.

Limitations

Because it is a strong base, great care needs to be taken to ensure that it is mixed rapidly into the formulation to avoid local increases in pH, which could have irreversible effects. Potassium hydroxide is available as both a solid and a concentrated liquid, both of which are very caustic, and precautions need to be taken to avoid these materials coming into contact with the skin. Once diluted in the food it is harmless.

Potassium hydroxide is a generally permitted additive under Directive 95/2/EC.

Typical Products

None known.

E526	Calcium hydroxide

Sources

Calcium hydroxide is made industrially by adding water to calcium oxide. It is commonly known as slaked lime. It is often made on site as part of the manufacturing process in which it is used.

Function in Food

Calcium hydroxide is a weak alkali and is used to lower acidity. It is used in the purification of sugar syrup.

Benefits

Calcium hydroxide is used in the purification of sugar because it neutralises the sugar syrup and is then removed by passing carbon dioxide into the mix, precipitating calcium carbonate and removing with it many of the organic colloidal impurities in the sugar.

It is used as a source of calcium ions to react with natural pectins in fruit to preserve the integrity of fruit particles in fruit pulp for jam manufacture.

Limitations

Calcium hydroxide is not very soluble in water.
It is a generally permitted additive under Directive 95/2/EC.

Typical Products

Used as a source of calcium for food fortification.

E527   Ammonium hydroxide

Sources

Ammonium hydroxide is prepared by passing ammonia gas into water.

Function in Food

Ammonium hydroxide is a base and is used to decrease the acidity of food formulations.

Limitations

Ammonium hydroxide is available only as a solution in water.
It is a generally permitted additive under Directive 95/2/EC.

Typical Products

None known.

E528   Magnesium hydroxide

Sources

Magnesium hydroxide is manufactured from the natural ore dolomite by heating and hydration, or is extracted from seawater.

Function in Food

Magnesium hydroxide has long been used as a pharmaceutical antacid, and in foods is used to reduce the acidity of products.

Benefits

It provides a source of magnesium, which is an essential mineral.

Limitations

Magnesium hydroxide is a generally permitted additive under Directive 95/2/EC.

Typical Products

Used as a source of magnesium for food fortification.

E529   Calcium oxide

Sources

Calcium oxide is made by heating limestone.

Function in Food

Calcium oxide is used as a source of calcium hydroxide (E526). It is also used as a dough conditioner in bread making and in the production of maize tortillas.

Limitations

Calcium oxide is a generally permitted additive under Directive 95/2/EC.

Typical Products

Used as a source of calcium for food fortification.

E530   Magnesium oxide

Sources

Magnesium oxide is made industrially by heating naturally occurring magnesium carbonate (dolomite).

Function in Food

Magnesium oxide is used as a source of magnesium hydroxide (E528).

Limitations

Magnesium oxide is a generally permitted additive under Directive 95/2/EC.

Typical Products
Used as a source of magnesium for food fortification.

E535	**Sodium ferrocyanide**
E536	**Potassium ferrocyanide**
E538	**Calcium ferrocyanide**

Sources
The ferrocyanides are made by reacting the respective metal cyanide with ferrous sulphate.

Function in Food
The ferrocyanides, particularly the sodium salt, are used as anticaking agents in table salt.

Limitations
The ferrocyanides are included in Annex IV of Directive 95/2/EC, where they are permitted only as an additive in salt and its substitutes and then only to a maximum level of 20 mg/kg.
The ferrocyanides have a joint ADI of 0.025 mg/kg body weight.

Typical Products
Table salt.

E541	**Sodium aluminium phosphate**

Sources
Sodium aluminium phosphate is a white, odourless powder made by the reaction of sodium hydroxide, aluminium oxide and phosphoric acid.

Function in Food
Sodium aluminium phosphate is an acidic product used as a raising agent with a carbon dioxide generator such as sodium bicarbonate (E500). It provides slow release of carbon dioxide and is used in commercial doughs and batters, where the dough is made up and is held refrigerated before cooking. Typically, about 20% of available carbon dioxide is released from the bicarbonate during mixing and the remainder is released during cooking.

It is also used in a mixture with monocalcium phosphate (E341) to provide release of carbon dioxide both before and during cooking.

Benefits

Sodium aluminium phosphate has a bland flavour and provides a uniform texture with a large bake-out volume. Its particular benefit is the low level of carbon dioxide released during refrigerated storage.

Limitations

Under Annex IV of Directive 95/2/EC sodium aluminium phosphate is permitted only to a level of 1 g/kg in bakery wares, specifically scones and sponges.

Typical Products

Scones and sponges.

E551	Silicon dioxide

Sources

Silicon dioxide (silica) is found in nature as sand. The products used in the food industry are synthetically produced amorphous silica. Two forms are available: silica aerogel, which is a microcellular silica, and hydrated silica, which is silicon dioxide, prepared by precipitation or gelling.

Function in Food

Food-grade silicon dioxide is an extremely fine powder with a very high ratio of surface area to weight. This allows it to coat the surface of powders to prevent them from sticking. This property also allows its use as a carrier and occasionally a thickener. A range of different silicas is available commercially with different properties.

Benefits

Silicon dioxide is used at very low levels to improve powder flow and processing. It can improve the performance of powders where the problem is caused by particle size distribution, fattiness or stickiness.

Limitations

Under Annex IV of Directive 95/2/EC as amended by Directives 98/72/EC, 2003/114/EC and 2006/52/EC, silicon dioxide and the silicates are free-flow agents permitted to *quantum satis* in a number of products and to a level of 10 g/kg in dried powdered foods (including sugars), salt and substitutes, sliced or grated hard, semi-hard and processed cheeses and analogues of these, 30 g/kg in seasonings and tin-greasing products and 50 g/kg in flavourings. There is no limitation in use (*quantum satis*) in dietary food supplements or food in tablet and coated tablet form. The limitations apply either individually or in combination with E552, E553a E553b, E554, E555, E556 and E559. It is limited in use as a carrier in emulsifiers and colours to a maximum of 5% and as a carrier in E171 (titanium dioxide) and E172 (iron oxides/hydroxides) to a maximum of 90% relative to the pigment. The powder is very dusty and care needs to be taken when handling it.

Typical Products
Powders for drinks and desserts, and sliced cheese.

E552	Calcium silicate

Sources
Calcium silicate is an amorphous powder prepared by a precipitation process from inorganic raw materials.

Function in Food
Calcium silicate is a fine white powder used as free-flow agent in food.

Benefits
Calcium silicate is used at low levels to improve the flow properties of powders. It is also used as an anti-tack agent to coat products that have a propensity to bond together.

Limitations
Under Annex IV of Directive 95/2/EC as amended by Directives 98/72/EC and 2006/52/EC, calcium silicate and other silicates are free-flow agents permitted to *quantum satis* in a number of products and to a level of 10 g/kg in dried powdered foods (including sugars), salt and substitutes, sliced or grated hard, semi-hard and processed cheeses and analogues of these, and 30 g/kg in seasonings and tin-greasing products. There is no limitation in use (*quantum satis*) in dietary food supplements or food in tablet and coated tablet form. The limitations apply either individually or in combination with E551, E553a E553b, E554, E555, E556 and E559. Calcium silicate is limited in use as a carrier in emulsifiers and colours to a maximum of 5%.

Typical Products
Emulsifiers, sliced cheese and powdered foods.

E553a   Magnesium silicate

Sources
Magnesium silicate is an amorphous powder prepared by a precipitation process from inorganic raw materials.

Function in Food
Fine white powder used as a free-flow agent in food.

Benefits
Magnesium silicate is used at low levels to improve the flow properties of powders. Its small particle size suits it for use where separation of particles that may bond together leads to improved flow characteristics.

Limitations
Under Annex IV of Directive 95/2/EC as amended by Directives 98/72/EC and 2006/52/EC, magnesium silicate and other silicates are free-flow agents permitted to *quantum satis* in a number of products and to a level of 10 g/kg in dried powdered foods (including sugars), salt and substitutes, sliced or grated hard, semi-hard and processed cheeses and analogues of these, and 30 g/kg in seasonings and tin-greasing products. There is no limitation in use (*quantum satis*) in dietary food supplements or food in tablet and coated tablet form. The limitations apply either individually or in combination with E551, E552, E553b, E554, E555, E556 and E559.

Typical Products
Powdered food ingredients.

E553b   Talc

Sources
Talc is a white to greyish powder. It is a naturally occurring form of magnesium silicate. Sources known to be associated with asbestiform minerals are not used as food-grade.

Function in Food

Talc is used as a dusting powder, anticaking agent and release agent. It is also used as a filtration aid.

Benefits

Talc is readily available, easy to handle and relatively inexpensive. Because it is relatively coarse, it is easier to handle than many anticaking agents.

Limitations

Talc is one of a number of anticaking agents that are permitted in Annex IV of Directive 95/2/EC, as amended by Directives 98/72/EC and 2006/52/EC, for a limited range of foods, whether alone or in combination with other silicates. It may be used to *quantum satis* in rice, chewing gum, foods in tablet form or dietary food supplements, but a maximum level of 10 g/kg applies in salt, dried powdered foods and sliced or grated hard, semi-hard and processed cheese and analogues of these. Silicates are also permitted to *quantum satis* for surface treatment of sausages and confectionery other than chocolate and to a maximum 30 g/kg in seasonings and tin-greasing products. The limitations apply either individually or in combination with E551, E552, E553a, E554, E555, E556 and E559.

Typical Products

Tablet coatings.

E554	Sodium aluminium silicate
E555	Potassium aluminium silicate
E556	Calcium aluminium silicate

Sources

There is no commercial preparation of calcium aluminium silicate. The other aluminium silicates are prepared by coprecipitation of soluble salts of aluminium and the appropriate metal. They are white, amorphous powders.

Function in Food

The aluminium silicates are fine powders that are used as free-flow agents in food powders. The use of sodium aluminium silicate is considerably greater than that of the potassium salt.

Potassium aluminium silicate is used as a carrier.

Benefits

The aluminium silicates are relatively inexpensive. Choices between free-flow agents will be affected by both improvement of powder flow and the effect on subsequent processing, as well as price. Potassium aluminium silicate is used where it is required to minimise the sodium content of the foodstuff.

Limitations

Under Annex IV of Directive 95/2/EC as amended by Directives 98/72/EC, 2003/114/EC and 2006/52/EC, silicon dioxide and the silicates are free-flow agents permitted to *quantum satis* in a number of products and to a level of 10 g/kg in dried powdered foods (including sugars), salt and substitutes, sliced or grated hard, semi-hard and processed cheeses and analogues of these, and 30 g/kg in seasonings and tin-greasing products. There is no limitation in use (*quantum satis*) in dietary food supplements or food in tablet and coated tablet form. The limitations apply either individually or in combination with E551, E552, E553a E553b and E559.

Potassium aluminium silicate (E555) is permitted as a carrier under Annex V where it is allowed in E171 titanium dioxide and E172 iron oxides and hydroxides and limited to a maximum of 90% relative to the pigment.

Typical Products

The aluminium silicates are used in drinks powders and desserts.

E558   Bentonite

Sources

Bentonite is a naturally occurring clay, which is extracted and then cleaned by washing before drying and milling.

Function in Food

Bentonite is used as a carrier for colours. It is also used widely as a filtration medium in the production of wine. For this latter use, it would be classified as a processing aid since it is not present in the finished product.

Benefits

Bentonite is an extremely absorbent material, which can be prepared with a high surface area per g.

Limitations

Under Directive 95/2/EC, bentonite is permitted only as a carrier for colours up to a maximum of 5%.

Typical Products

None known.

E559	Kaolin (aluminium silicate)

Sources

Kaolin is a naturally occurring white clay, which is mined and washed before drying and milling.

Function in Food

Kaolin is a fine powder used as a dusting powder, anti-caking agent and release agent. It is also used as a filtration aid.

Benefits

Kaolin is readily available, easy to handle and relatively inexpensive.

Limitations

Under Annex IV of Directive 95/2/EC as amended by Directives 98/72/EC and 2006/52/EC, kaolin and other silicates are free-flow agents permitted *quantum satis* in a number of products and to a level of 10 g/kg in dried powdered foods (including sugars), salt and substitutes, sliced or grated hard, semi-hard and processed cheeses and analogues of these, and 30 g/kg in seasonings and tin-greasing products. There is no limitation in use (*quantum satis*) in dietary food supplements or food in tablet and coated tablet form. The limitations apply either individually or in combination with E551, E552, E553a, E553b, E554, E555 and E556.

Typical Products

None known.

E570   Fatty acids

Sources

    The fatty acids include stearic, palmitic and oleic acids. They are made by fractionation of natural fats such as tallow followed by acidification. They can be used alone or in mixtures.

Function in Food

    Fatty acids have a number of functions, including plasticisers for chewing gum and anitfoaming agents for jams.

Limitations

    Fatty acids are generally permitted under Annex I of Directive 95/2/EC.

Typical Products

    Chewing gum.

E574   Gluconic acid

Sources

    Gluconic acid is an organic acid occurring naturally in plants, fruits and other foodstuffs such as wine and honey. The material of commerce is prepared by fermentation of glucose, which is itself produced by enzymic hydrolysis of starch. The process results in a mixture of both gluconic acid and glucono-delta-lactone (E575). The two products are separated by the crystallisation of the lactone. The acid is supplied as a 50% solution since dehydration leads to the formation of the lactone.

Function in Food

    Gluconic acid is used as a mild acid and as a chelating agent, particularly in foods with a neutral pH. However, its role as a metal chelation agent finds its major applications outside the food industry.

Benefits

    The use of gluconic acid allows the reduction of pH in foods where the acid taste of citric, malic or even lactic acid would not be acceptable. The chelating effect is used to bind metal ions that might otherwise catalyse oxidation reactions that would decrease shelf-life.

Limitations

Gluconic acid is completely metabolised in the body in the same way as a carbohydrate. Under Annex I of Directive 95/2/EC, gluconic acid is a generally permitted additive.

Typical Products

Soft drinks, confectionery and fruit preparations.

E575	Glucono-delta-lactone (GdL)

Sources

Glucono-delta-lactone is a neutral cyclic ester of gluconic acid, produced with gluconic acid by the fermentation of glucose or glucose-containing raw materials such as glucose syrup. It is separated from the acid by crystallisation and the material of commerce is a white crystalline powder or granule.

Function in Food

When dissolved in water, GdL hydrolyses to gluconic acid, so it is used as a slow-release acidifier, for curdling milk proteins in cheese and soya proteins in tofu manufacture, and in the ripening process of a wide range of sausages. In bakery, the released gluconic acid serves as a chemical leavening agent by reacting with sodium bicarbonate to give carbon dioxide, and in other products it acts as part of the preservation system. The chelating ability of the acid is particularly useful where iron and copper are present.

Benefits

The principal benefit of GdL is the slow and steady release of the acid, which lies at the heart of most of the applications. In precipitation of proteins, the steady reduction of pH has a number of benefits in product quality and reduction of production times, and the mild flavour is more compatible with the final product taste than is the case with most organic acids. In some cases it is used to wholly or partially replace bacterial starter cultures. It has a similar role in the processing of meat products, where it may also allow a reduction in the nitrite usage. In bakery products, mixtures of lactone with sodium bicarbonate release carbon dioxide slowly and continuously, in a comparable way to yeast, but often in a shorter time. It has advantages over some other leavening agents, being less sharply acid and having no bitter or soapy aftertaste. In gelling mixtures with alginate, the steady decrease in pH improves the final gel texture.

The chelation of iron, copper and other metal ions inhibits oxidation reactions and extends shelf-life.

Limitations

GdL is completely metabolised in the body in the same way as a carbohydrate. Under Directive 95/2/EC, it is a generally permitted additive and is permitted in Annex II for ripened cheese, Mozzarella cheese, fresh pasta and canned and bottled fruit and vegetables.

GdL must be kept cool and dry as exposure to moisture will initiate the hydrolysis to the acid and lumping of the product.

Typical Products

Cottage cheese, Feta-type cheese, Mozzarella, tofu, raw and cooked sausages, refrigerated or frozen prepared doughs, instant dough mixes and canned seafood.

E576	**Sodium gluconate**
E577	**Potassium gluconate**
E578	**Calcium gluconate**
E579	**Ferrous gluconate**

Sources

Glucono-delta-lactone and gluconic acid are produced together by the fermentation of glucose. Sodium gluconate is made by neutralising the mixture with sodium hydroxide.

Potassium and calcium gluconates are made by neutralising gluconic acid with the appropriate hydroxide or carbonate.

Ferrous gluconate is made either by neutralising gluconic acid with ferrous carbonate or by reacting calcium gluconate with ferrous sulphate.

The sodium, potassium and calcium salts are white powders or crystals, which are soluble in water.

Ferrous gluconate is a pale yellowish grey or green powder, which is soluble in water.

Function in Food

The major uses of sodium gluconate are outside the food industry. Its ability to complex metal ions leads to extensive uses in cleaning products and in the construction industry. In the food industry, its main use derives from its property of covering or reducing the bitterness of other ingredients. Thus it is used in low-sugar products to cover the bitter or metallic taste of saccharin.

Potassium and calcium gluconates are used mainly for fortification, but the calcium salt is also used as a readily soluble form of calcium, for example in the setting of dessert mixes or the precipitation of proteins. Potassium gluconate is used as a buffer in soft drinks and as a component of salt replacers. The ferrous salt finds its principal application as a pro-oxidant in darkening green olives by an oxidation reaction.

Benefits

The gluconates have a number of advantages over other salts used for fortification. These include high bioavailability, good water solubility and neutral taste.

Limitations

It is important to note that vitamins and minerals added to fortify foods are covered by EC Regulation 1925/2006. As additives, sodium, potassium and calcium gluconate are permitted under Annex I of Directive 95/2/EC as generally permitted additives. Because of its iron content, ferrous gluconate is in Annex IV of the Directive, where it is permitted only for use in olives darkened by oxidation to a maximum of 150 mg/kg.

Typical Products

Sodium gluconate in low-calorie chewing gum.
Potassium gluconate in baked goods, milk drinks, sport and health drinks, and nutritional bars.
Calcium gluconate in drinks, desserts and dairy products.
Ferrous gluconate in darkened olives.

E585   Ferrous lactate

Sources

Ferrous lactate is made by reacting ferrous sulphate with calcium or sodium lactate.

Function in Food

Ferrous lactate is used to preserve and darken the colour of olives.

Limitations

Ferrous lactate becomes darker and less soluble in water on exposure to air, so it must be kept, as a raw material, in a tightly sealed opaque container.

Under Directive 95/2/EC, ferrous lactate is permitted only in olives darkened by oxidation to a maximum of 150 mg/kg.

E586	4-Hexylresorcinol

Sources

Hexylresorcinol is a chemically synthethised compound with anaesthetic, antiseptic and antihelmintic properties. It is freely soluble in ether and in acetone and very slightly soluble in water.

Function in Food

Antioxidant.

Benefits

The use of 4-Hexylresorcinol can help prevent melanosis (black spots) forming in the shell of raw, refrigerated and frozen crustaceans within a few hours of them being harvested.

Limitations

It is permitted under part D of Annex III to Directive 95/2/EC, as amended by Directive 2006/52/EC, where it is permitted in fresh, frozen and deep-frozen crustaceans up to 2 mg/kg as residues in crustacean meat.

Typical Products

Crustacean meat.

E620	Glutamic acid
E621	Monosodium glutamate
E622	Monopotassium glutamate
E623	Calcium diglutamate
E624	Monoammonium glutamate
E625	Magnesium diglutamate

Sources

Glutamic acid is an amino acid abundant in nature, either alone or as a component of proteins. As an individual amino acid it is present in tomatoes and seaweed. Of its salts, monosodium glutamate (MSG) is the only one used to any significant extent in the food industry. MSG is made by fermentation, usually starting from starch or molasses. The product of fermentation is separated by filtration, dissolved and neutralised with an alkaline sodium salt such as sodium hydroxide.

All the forms of glutamic acid and its salts are commonly referred to in the industry as glutamate.

Function in Food

Glutamate is used to develop and enhance the flavour of, mainly, savoury products. It also has its own characteristic flavour, which is considered by some people to be a fifth basic taste, "umami", in addition to the original four of sweet, salt, sour and bitter.

Benefits

Glutamate works in a wide variety of dishes, strengthening, developing and rounding savoury flavour.

Limitations

The glutamates are permitted in Annex IV of Directive 95/2/EC in foods in general to a maximum of 10 g/kg. Some exceptional foods, Parmesan cheese for example, naturally contain glutamate higher than this limit. They are also permitted in seasonings and condiments to *quantum satis*.

The taste of MSG has a self-limiting characteristic. Once the correct amount has been used, any additional quantity contributes little, if anything. Indeed excess MSG can lead to a decrease in palatability.

Typical Products

Soups, sauces, prepared meals and sausages.

E626	Guanylic acid
E628	Dipotassium guanylate
E629	Calcium guanylate
E630	Inosinic acid
E632	Dipotassium inosinate
E633	Calcium inosinate

None of the above additives is produced commercially or actually used in the food industry. Each of them is deemed to function similarly to its sodium salt counterpart.

| E627 | Disodium 5'-guanylate (disodium guanosine 5'-monophosphate, GMP) |

Sources

Disodium guanylate is produced by the following methods:

1) Enzymic hydrolysis of yeast ribonucleic acid (RNA), followed by removal of other nucleotides and neutralisation by sodium hydroxide.
2) Fermentation of a sugar source into guanosine, followed by phosphorylation and neutralisation steps.
3) Fermentation of a sugar source into guanylic acid, followed by neutralisation.

Function in Food

Guanylate is the substance responsible for the flavour enhancement function of shiitake mushrooms. Disodium guanylate is commonly added to food products as a 1:1 mixture with disodium inosinate.

Benefits

The flavour enhancement function of disodium guanylate is synergistically increased when combined with a glutamate source such as monosodium glutamate. The flavour of disodium guanylate is 2.4–3 times stronger than that of disodium inosinate. In addition to enhancing savoury flavour, disodium guanylate also smoothes acidity and saltiness, suppresses bitterness and metallic notes, and masks off-flavours of protein hydrolysates and yeast extracts.

Limitations

Disodium guanylate is permitted in Annex IV of Directive 95/2/EC in foods in general to a maximum of 500 mg/kg and in seasonings and condiments to *quantum satis*.

Typical Products

Soups, sauces, processed meat, poultry and seafood.

E631	Disodium 5'-inosinate (disodium inosine 5'-monophosphate, IMP)

Sources

Disodium inosinate is produced by the following methods:

1) Enzymic hydrolysis of yeast ribonucleic acid (RNA), followed by removal of other nucleotides and neutralisation by sodium hydroxide.
2) Fermentation of a sugar source into inosine, followed by phosphorylation and neutralisation steps.
3) Fermentation of a sugar source into inosinic acid, followed by neutralisation.

Function in Food

Inosinate is naturally present in protein foods. It is used to enhance the flavour of red meat, poultry and seafood. Simply adding more meat may not be as effective in increasing the flavour as the concentration of flavour substances is relatively low and they are released only slowly from the tissue of the meat by chewing. Disodium inosinate is commonly added to food products by itself or as a 1:1 mixture with disodium guanylate.

Benefits

The flavour enhancement function of disodium inosinate is synergistically enhanced when combined with glutamate sources such as monosodium glutamate. Disodium inosinate is a less effective enhancer than the disodium 5'-ribonucleotides and disodium guanylate, being 0.33–0.44 times as strong as the guanylate. In addition to enhancing savoury flavour, disodium inosinate also smoothes acidity and saltiness, suppresses bitterness and metallic notes, and masks off-flavours of protein hydrolysates and yeast extracts.

Limitations

Disodium inosinate is permitted in Annex IV of Directive 95/2/EC in foods in general but to a maximum of 500 mg/kg and in seasonings and condiments to *quantum satis.*

Typical Products

Soups, sauces, processed meat and seafood.

E634   Calcium 5'-ribonucleotides

Sources

Calcium 5'-ribonucleotides are obtained by reacting disodium 5'-ribonucleotides and calcium chloride.

Function in Food

Calcium 5'-ribonucleotides comprise approximately a 1:1 mixture of calcium 5'-guanylate and calcium 5'-inosinate. They are added to food products as flavour enhancers.

Benefits

The flavour enhancement function of calcium ribonucleotides is synergistically heightened when combined with glutamate sources such as monosodium glutamate. The enhancing ability of calcium ribonucleotides is equal to that of disodium 5'-ribonucleotides. They are far less soluble in water than sodium ribonucleotides, and are therefore less susceptible to phosphatase enzymes, which remove the phosphate moiety from the nucleotide molecule. Other functions of the additive include smoothing sharp acidity and saltiness, suppression of bitterness and metallic notes and masking of off-flavours of protein hydrolysates and yeast extracts.

Limitations

Calcium 5'-ribonucleotides are permitted in Annex IV of Directive 95/2/EC in foods in general to a maximum of 500 mg/kg and in seasonings and condiments to *quantum satis.*

Typical Products

The usage of calcium ribonucleotides is very limited in Europe. They are mainly used as flavour enhancers in processed seafood, processed meat and poultry, and fermented soya-bean soup (miso) bases.

E635   Disodium 5'-ribonucleotides (sodium 5'-ribonucleotides)

Sources

Disodium 5'-ribonucleotides are produced by the following methods:

1) Enzymic hydrolysis of yeast ribonucleic acid (RNA), followed by removal of other nucleotides and neutralisation by sodium hydroxide.

2) Fermentation of a sugar source into guanosine and inosine, followed by phosphorylation and neutralisation steps.

3) Separate production of disodium 5'-guanylate and disodium 5'-inosinate, followed by a mixing process.

Function in Food

Disodium 5'-ribonucleotides comprise approximately a 1:1 mixture of disodium 5'-guanylate and disodium 5'-inosinate. They are the most commonly used form of nucleotide flavour enhancers.

Benefits

The flavour enhancement function of disodium 5'-ribonucleotides is synergistically heightened when combined with glutamate sources such as monosodium glutamate. Typically, they are combined with MSG at a ratio of between 2:98 and 10:90. The relative intensity of disodium 5'-ribonucleotides is 1.65–2 times that of disodium inosinate. Other functions of the additive includes smoothing sharp acidity and saltiness, suppression of bitterness and metallic notes and masking of off-flavours of protein hydrolysates and yeast extracts.

Limitations

Sodium 5'-ribonucleotides are permitted in Annex IV of Directive 95/2/EC in foods in general to a maximum of 500 mg/kg and in seasonings and condiments to *quantum satis*.

Typical Products

Disodium 5'-ribonucleotides are commonly used as flavour enhancers in soups, sauces, snack seasonings, processed meat and poultry, processed seafood, cheese products, tomato-based products and other common processed foods.

E640   Glycine

Sources

Glycine is a naturally occurring amino acid that is a part of most proteins. The material of commerce is produced synthetically.

Function in Food

Glycine is used in the food industry as a preservative, an antioxidant, and a browning and seasoning agent. Glycine is required by the body for the maintenance of the central nervous system and immune system.

Benefits

Glycine has a naturally sweet taste and is used alone or as an enhancer of savoury flavours. It can also enhance the taste of saccharin and mask the bitter aftertaste of intense sweeteners.

The Maillard reaction between sugars and amino acids is the reaction that produces browning and flavour development in roasted and baked products from roast meat to cakes. The presence of free amino acids increases the rate of this reaction and glycine is used for this purpose.

Glycine is also used to chelate metal ions that would otherwise catalyse autooxidation reactions, and it has an inhibitory effect on bacteria.

Limitations

Glycine and its sodium salt are generally permitted additives in Annex I of Directive 95/2/EC. Glycine shows little preservative activity on moulds and yeast.

Typical Products

Meat products and dietetic foods. It is also used extensively in other industries.

E650   Zinc acetate

Sources

Zinc acetate is also known as acetic acid, zinc salt and zinc diacetate. Zinc is an essential trace element for animals and humans and occurs mainly in meat and seafood. It is an integral component of many metalloenzymes.

Function in Food

Zinc acetate is added to chewing gum as a flavour enhancer. It is added to provide an astringent taste, and particularly to intensify the taste of bitterness from ingredients such as coffee or grapefruit.

Limitations

According to Directive 95/2/EC, as amended by Directive 2001/5/EC, zinc acetate is permitted only in chewing gum to a maximum 1000 mg/kg.

E900	Dimethyl polysiloxane (silicone, silicone oil, dimethyl silicone)

Sources

Dimethyl polysiloxane is made from silica and oil-derived chemicals.

Function in Food

Dimethyl polysiloxane is a surfactant. In foams it occupies the air-water interface and can act in two ways. At low concentration it is a foam stabiliser and at higher concentrations it causes foams to collapse.

Benefits

Dimethyl polysiloxane is mainly used as an antifoaming agent but is also used to stop hot liquids from sticking to equipment. It is useful in products such as jam, preventing frothing when boiling, and in carbonated drinks, which tend to froth when being filled into bottles or cans.

Limitations

According to Directive 95/2/EC as amended by Directives 98/72/EC and 2003/114/EC, dimethyl polysiloxane is permitted in a limited range of products, such as jam, soups, frying oils, drinks, canned and bottled fruit, flavourings to a maximum of 10 mg/kg and in chewing gum to 100 mg/kg.

It is permitted in cider (excluding *cidre bouché*) to 10 mg/l and in Annex V as a carrier of glazing agents for fruit.

It has an ADI of 1.5 mg/kg body weight.

Typical Products

Dimethyl polysiloxane is used in soft drinks and catering cooking oil.

E901   Beeswax

Sources
Beeswax is purified from naturally produced honeycomb.

Function in Food
Beeswax is used to provide a surface finish to stop sticky items sticking together and also to impart a shine to the surface of products.

Benefits
Beeswax is a soft wax, which allows it to be easily applied to soft sticky items such as gums. To panned goods it gives an instant shine, but, being a soft wax, this does not last a long time. It is used in blends with other, harder waxes to make them easier to apply and to give good adhesion to the piece. Beeswax has the best taste of all the waxes.

Limitations
Under Directives 95/2/EC, as amended by Directives 98/72/EC and 2006/52/EC, beeswax is permitted only as a glazing agent in a range of snacks, confectionery and bakery items, and food supplements, and on the surface of some fresh fruit.

Typical Products
Small confectionery items such as jellies and gums, and tablets.

E902   Candelilla wax

Sources
Candelilla wax is a natural wax extracted from the leaves of the candelilla plant, *Euphorbia antisyphilitica* Zucc.

Function in Food
Candelilla was the original wax used to plasticise chewing gum. It is now used to impart a shine to small pieces of confectionery and to coat the surface of fruit to reduce drying during storage.

Benefits

Candelilla wax is softer and thus easier to apply than carnauba wax. Because it blends with oils, it aids the retention of flavours in gum. Candelilla wax is the best moisture barrier of the natural waxes.

Limitations

Under Directive 95/2/EC, as amended by Directives 98/72/EC and 2006/52/EC, candelilla wax is permitted only as a glazing agent in a range of snacks, confectionery and bakery items and food supplements, and on the surface of some fresh fruit.

Typical Products

Chewing gum and small pieces of confectionery.

E903	Carnauba wax

Sources

Carnauba wax is a natural material extracted from the fronds of the Brazilian wax palm, *Copernicia prunifera* (Mill.).

Function in Food

Carnauba wax is used to provide a shine to the surface of small sweets. It can also be used to coat the surface of fruit to reduce drying during storage.

Benefits

Carnauba wax is the hardest known wax, with a melting point of 78–88 °C. Hard waxes are better for keeping a shine on products throughout the shelf-life.

Limitations

Under Annex IV of Directive 95/2/EC as amended by Directives 98/72/EC and 2003/114/EC, carnauba wax is permitted only as a glazing agent in a range of snacks, confectionery and bakery items and food supplements, and on the surface of some fresh fruit up to specified limits.

Carnauba wax has an ADI of 0-7 mg/kg body weight.

Carnauba wax works well where pressure can be applied to produce the shine. This makes it inappropriate for delicate products.

Typical Products

Carnauba wax is used to glaze small pieces of chocolate confectionery, gums, jellies and chewing gum

---

E904   Shellac

Sources

Shellac is purified and refined from a resinous secretion of an Indian scale insect, *Laccifer lacca.*

Function in Food

Shellac has a number of uses. It is used to provide a polished surface on products, to prevent sticky items from sticking together, and to reduce the moisture loss of fresh fruit.

Benefits

Shellac is insoluble but dispersible in water and provides high gloss without the need for pressure. It can be used in combination with other glazing agents

Limitations

Under Annex IV of Directive 95/2/EC, as amended by Directive 2006/52/EC, shellac is permitted only as a glazing agent in a range of snacks, confectionery and bakery items, food supplements, and on the surface of fresh fruit. No limits are set for the applications.

Typical Products

Shellac is used on small coated pieces, such as chocolate and flavoured coated confectionery, biscuits, and gums.

E905   Microcrystalline wax

Sources

Microcrystalline wax is a hydrocarbon wax isolated and purified in the petroleum industry.

Function in Food

Microcrystalline wax is a chemically inert wax that is used as a lubricant for chewing gum and to stop sticky products sticking together.

Limitations

Microcrystalline wax is permitted under Directive 98/72/EC only for the surface treatment of confectionery (excluding chocolate), chewing gum and certain fruit.

E907	Hydrogenated poly-1-decene

Sources

Hydrogenated poly-1-decene is prepared by the hydrogenation of mixtures of trimers, tetramers, pentamers and hexamers of 1-decenes. Pure 1-decene is itself made from ethylene. Minor amounts of molecules with carbon numbers less than 30 may be present. It is insoluble in water, slightly soluble in ethanol and soluble in toluene.

Function in Food

Glazing agent.

Benefits

It is thermally and microbiologically very stable, with very low volatility since the boiling range starts only at 320 °C.

Limitations

It is a permitted as a glazing agent under Annex IV of Directive 95/2/EC as amended by 2003/114/EC, for use in a limited number of foodstuffs, namely sugar confectionery and dried fruits up to a maximum level of 2g/kg.

The acceptable daily intake (ADI) for hydrogenated poly-1-decene is given by the Joint FAO/WHO Expert Committee on Food Additives as being between 0-6 mg/kg body weight

Typical Products

Sugar confectionery, dried fruits

E912	Montan acid esters

Sources

Montan wax is a hard brown wax obtained by solvent extraction of a fossilised vegetable wax. It is refined, oxidised with chromic acid, and esterified to give a pale yellow wax.

Function in Food

Montan wax is used to provide a protective layer on the skins of fruit to reduce loss of moisture during storage. The treatment is applied only to fruit where the skin is not eaten.

Limitations

Under Directives 95/2/EC and 98/72/EC, montan acid esters are permitted only for the surface treatment of fresh citrus and exotic fruit.

Montan wax has an unpleasant taste.

E914	Oxidised polyethylene wax

Sources

Oxidised polyethylene wax is prepared by the mild air oxidation of polyethylene.

Function in Food

Oxidised polyethylene wax is used to provide a protective coating to fresh fruit to reduce moisture loss during storage.

During the washing process, fruit tends to lose some of its natural wax, and glazing agents such as oxidised polyethylene wax are used to replace it. The wax is sprayed on as a very small droplet size emulsion.

Limitations

Under Directives 95/2/EC and 98/72/EC, oxidised polyethylene wax is permitted only for the surface treatment of fresh citrus and exotic fruit.

E920   L-cysteine hydrochloride

Sources
L-cysteine is a high-sulphur-containing amino acid synthesised by the liver. Industrially it is made by extraction or fermentation of human hair, chicken or duck feathers, or pig bristle.

Function in Food
L-cysteine is a flour-treatment agent, which is used to help the gluten relax in doughs that are heavily manipulated, such as pizza bases.

Benefits
The major uses of L-cysteine in the food industry are in flavourings, where it is an important ingredient in recipes for Maillard flavours, and in the fortification of functional foods. A number of health benefits are claimed.

Limitations
L-cysteine is permitted only as an additive under Directive 98/72/EC where it is permitted only as a flour-improvement agent. Uses in flavourings and fortification are covered by other legislation. L-cysteine may interfere with insulin metabolism. Diabetics are therefore advised not to use L-cysteine supplements without consulting their physician.

Typical Products
Pizza bases, hamburger buns, pet food and dietary supplements.

E927b   Carbamide

Sources
Carbamide is white crystalline material made by reaction of carbon dioxide with ammonia followed by purification. It is produced in large volumes for agricultural and pharmaceutical use.

Function in Food
When foods containing fermentable carbohydrate, such as starch or sugars, are consumed, bacteria in the mouth use some of the carbohydrate and convert it to acid. When the saliva becomes acid, the teeth lose enamel and cavities can begin to form. Chewing gum is often used after meals to generate a

flow of saliva and, in so doing, to decrease the acidity of the saliva and thus the rate of development of dental caries. Carbamide is included in chewing gum to assist in this process of decreasing the acidity of saliva.

Limitations

According to Directive 95/2/EC, carbamide is permitted only in chewing gum without added sugar.

E938   Argon

Sources

The components of air with approximate ratios are shown below:
78.1%   nitrogen
20.9%   oxygen
0.9%   argon
0.1%   carbon dioxide, rare gases, moisture

To produce argon, air is filtered, dried and compressed. The compressed air is then expanded to produce cold, which liquefies the air. This liquid is introduced to distillation columns, where separation takes place. The argon can be withdrawn as a gas or cryogenic liquid. Purity levels are typically better than 99.995%.

Function in Food

Argon is used to replace air within food packages in a technique known as modified-atmosphere packaging. In this system, argon may be used as an alternative to nitrogen, where a physically inert atmosphere is required, and where some biochemical properties of argon are seen to be beneficial in terms of colour and flavour retention, and where some enzyme systems associated with spoilage in specific food groups may be modified or retarded.

It can be used advantageously for blanketing and sparging liquids, owing to its inert and other physical properties.

Benefits

Food product shelf-life can be extended considerably compared with products stored in air.

Limitations

Argon is a generally permitted additive in Annex I of Directive 95/2/EC.

Argon occurs in the atmosphere at 0.94% by volume, so is relatively rare. This means that its cost, compared with that of other packaging gases, is high, so it must be used in specific cases where there are justifiable grounds from a cost-benefit point of view. Physically, it is a denser, more soluble gas than nitrogen, so in some situations gas usage can be reduced compared with nitrogen. Being denser than air, consideration must be given to safety aspects in confined or low-lying working environments.

Typical Products

Snack foods, cooked meats, pizzas, recipe dishes and wine.

E939   Helium

Sources

The major source of helium is from natural gas wells. The gas is obtained from a liquefaction and stripping operation.

Function in Food

There are no known commercial applications for helium in food processing or packaging.

Limitations

Helium is a generally permitted additive under Annex I of Directive 95/2/EC.

E941   Nitrogen

Sources

The components of air with approximate ratios are shown below:
78.1%   nitrogen
20.9%   oxygen
0.9%   argon
0.1%   carbon dioxide, rare gases, moisture

To produce nitrogen, air is filtered, dried and compressed. The compressed air is then expanded to cool it, which liquefies the air. This liquid is introduced to distillation columns, where separation takes place. The nitrogen can be withdrawn as a gas or cryogenic liquid. Purity levels are typically better than 99.995%.

Function in Food

In modified-atmosphere packaging, nitrogen is introduced into a food package to replace air, as an inert packaging gas. Air contains 20.9% oxygen, and it is this oxygen that can be detrimental to the quality shelf-life of a range of food products, through a process known as oxidative rancidity. This applies especially to products that have an inherent fat content or an oily coating, such as those that have been fried.

The application of an inert nitrogen atmosphere will displace oxygen and discourage the growth of aerobic spoilage bacteria.

Nitrogen has low solubility in water and fats, and so it is often used in combination with carbon dioxide where absorption of carbon dioxide could cause package collapse.

Nitrogen can be used to replace air in the whipping and aeration of creams and mousses in the dairy industry.

Nitrogen is used in the food industry for rendering foodstuffs inert and sparging liquids and oils to remove dissolved oxygen. Liquid nitrogen is used widely in food freezing and chilling.

Benefits

Food product shelf-life can be extended considerably by the exclusion of oxygen.

Limitations

Nitrogen is a generally permitted additive under Annex I of Directive 95/2/EC. Being inert and non-reactive, nitrogen has little bacteriostatic effect when used as a single packaging gas, so it is best used where products are initially bacteriologically clean or on its own with low-water-content products.

Typical Products

Dried foods, snack foods, dried milk and potato powder.

E942	Nitrous oxide

Sources

Nitrous oxide is most commonly obtained by the thermal decomposition of ammonium nitrate. It may also be obtained by controlled reduction of nitrates or nitrites and the thermal decomposition of hydroxylamine.

Function in Food

In modified-atmosphere packaging, nitrous oxide is a permitted packaging gas, but is not used for general food-packaging applications. It is used in the dairy industry for several applications, where it has the property of being able to reduce the oxidation of lipids by any residual air.

It is used in whipping creams and mousses and as a propellant in aerosol creams.

Benefits

Shelf-life in packaged dairy products can be extended by the use of nitrous oxide, where oxidative rancidity can be avoided.

Limitations

Nitrous oxide is a generally permitted additive under Annex I of Directive 95/2/EC. Nitrous oxide is a very reactive gas, requiring similar safety measures to oxygen.

Typical Products

Ready-to-serve whipped cream.

E943a Butane
E943b Iso-butane

Sources

Butane (also known as n-butane) and iso-butane are obtained from natural gas by fractionation. They are colourless, odourless, flammable gases at normal temperatures and pressures. They are readily liquefied under pressure at room temperature and are stored and shipped as liquids.

Function in Food

Butane and iso-butane can be used as propellants in vegetable oil pan sprays and water-based emulsion sprays.

Limitations

According to Directive 95/2/EC, as amended by Directive 2001/5/EC, butane is permitted only in vegetable oil pan sprays (for professional use only) and in water-based emulsion sprays in accordance with the principle of '*quantum satis*'.

E944	Propane

Sources

Propane (also known as dimethylmethane) is obtained from natural gas by fractionation. It is a colourless, odourless, flammable gas at normal temperatures and pressures, which is easily liquefied under pressure at room temperature. Propane is stored and shipped in the liquid state.

Function in Food

Propane can be used as a propellant in vegetable oil pan sprays and water-based emulsion sprays.

Limitations

Under the provisions of Directive 95/2/EC, as amended by Directive 2001/5/EC, propane may be used only in vegetable oil pan sprays (for professional use only) and in water-based emulsion sprays in accordance with the principle of '*quantum satis*'.

E948	Oxygen

Sources

The components of air with approximate ratios are shown below:
78.1% nitrogen
20.9% oxygen
0.9%   argon
0.1%   carbon dioxide, rare gases, moisture

To produce oxygen, air is filtered, dried and compressed. The compressed air is then expanded to cool it, which liquefies the air. This liquid is introduced to distillation columns, where separation takes place. The oxygen can be withdrawn as a gas or cryogenic liquid. Purity levels are typically better than 99.995%.

Function in Food

In modified-atmosphere packaging, the objective is generally to exclude oxygen from a food package as it is this oxygen (from air) that is used metabolically by aerobic spoilage microorganisms, and which also causes rancidity in a range of food products with a fat component (e.g. dairy products).

The exception to this is where oxygen is incorporated into a modified-atmosphere pack for the following reasons:

- In a high concentration, combined with carbon dioxide, to maintain oxygenation of myoglobin in fresh meats for bloom (a typical mixture might be 80% oxygen, 20% carbon dioxide).
- To produce a low-oxygen environment (of typically 5%) in combination with nitrogen to control and reduce the respiration rate of cut, prepared produce.
- In combination with carbon dioxide and nitrogen, to maintain an aerobic atmosphere in the packaging of white fish species.

Research work has shown that there may be some benefit in the use of high-oxygen atmospheres in the packaging of some species of respiring produce.

Benefits

Modified-atmosphere packaging is used to increase the quality shelf-life of a food product, and to enhance the presentation of the product compared with the same product offered in contact with air.

In fresh meats packaged in a mixture of oxygen and carbon dioxide, the colour presentation is considerably enhanced compared with that in air, and microbial spoilage is reduced by the presence of carbon dioxide.

For packaging respiring produce, it is advantageous to reduce the respiration rate of the product, thereby reducing the rate of deterioration. The low-oxygen environment is maintained by matching the rate of oxygen reduction with a packaging film of similar permeability, to maintain an equilibrium modified atmosphere.

Oxygen is incorporated into some fish packs to maintain an aerobic atmosphere, to avoid conditions favourable to the growth of anaerobic pathogens such as *Clostridium botulinum*.

Limitations

Oxygen is a generally permitted additive under Annex I of Directive 95/2/EC.

Oxygen is used in modified-atmosphere packaging selectively and specifically, as outlined above, and the effect of its use, in preferentially allowing conditions favourable to aerobic bacteria, must be understood in conjunction with other food preservation hurdles such as storage temperature.

When using high-oxygen mixtures (greater than 25%), it is important that packaging machinery be fully oxygen-compatible. Gas using and distribution equipment must also be oxygen-compatible and care must be taken in terms of system operational safety.

Typical Products

Fresh meat, cut and prepared lettuce, white fish and seafood.

E949	Hydrogen

Sources

Hydrogen is produced by catalytically reforming natural gas or other hydrocarbon fuels with superheated steam at elevated temperatures. It can also be produced by electrolytic decomposition of alkalised water, and by the cracking of gaseous ammonia.

Function in Food

The principal use of hydrogen is in fats and oils processing. Used with a catalyst, such as nickel, hydrogen converts the double bonds in unsaturated oils to single bonds creating stable, saturated fats. It changes liquid oils to semi-solid, hard or plastic fats. Hydrogen has also been considered as an innovative component of gas mixtures used for modified atmosphere packaging applications.

Limitations

Hydrogen is a generally permitted additive under Annex I of Directive 95/2/EC, as amended by Directive 2001/5/EC.

E950	Acesulfame K

Sources

Acesulfame K belongs to the group of oxathiazinone dioxide sweeteners. It is the potassium salt of 6-methyl-1,2,3-oxathiazine-4(3H)-one-2,2-dioxide. Several synthesis routes for acesulfame K have been described.

Function in Food

Acesulfame K is a non-cariogenic, non-laxative intense sweetener used in a wide range of foods, including foods for diabetics. Since it is not metabolised by the human body, it passes through the digestive system unchanged and is, therefore, completely "calorie-free". The sweetness potency of acesulfame K is approximately 200 times that of sucrose. Additionally, it can act as a flavour enhancer.

Benefits

Acesulfame K provides a clean sweetness with a fast onset. With other intense sweeteners, acesulfame K shows synergistic effects, which leads to a more sugar-like taste and additional sweetness enhancement. In particular, blends of acesulfame K and aspartame have a high synergy. Other synergistic taste enhancements have been demonstrated in blends of acesulfame K and alitame, cyclamate, neohesperidine DC (NHDC) and sucralose.

Acesulfame K is stable to a wide range of processing conditions, tolerating pH levels from 3 to 9 and temperatures up to 200 °C. It is highly soluble in water.

Limitations

Acesulfame K is permitted under the European Sweetener Directive 94/35/EC and its amendments 96/83/EC and 2003/115/EC for use in more than 40 food applications.

When used in high concentrations above normal use levels, acesulfame K may have a slight aftertaste. The Acceptable Daily Intake (ADI) for acesulfame K was set at 15 mg/kg body weight according to JECFA. Typical usage levels are 100–300 mg/litre in beverages and up to 2,000 mg/kg in confectionery and baked goods. Acesulfame K is approved in more than 90 countries.

Typical Products

Acesulfame K is used in all fields of application of intense sweeteners. Key areas of application are beverages, yoghurts, ice cream and other dairy products, desserts, confectionery, chewing gum, pharmaceutical products and table-top sweeteners.

E951	Aspartame

Sources

Aspartame (N-L-α-aspartyl-L-phenylalanine-methylester) is the methyl ester of a dipeptide composed of the amino acids L-aspartic acid and L-phenylalanine. The production of aspartame normally starts from L-phenylalanine or L-phenylalanine methylester and L-aspartic acid. The production follows the common routes of peptide synthesis, as the L-configuration of the amino acids has to be retained. An alternative to chemical peptide synthesis is enzymic formation of the peptide bond; both processes are used commercially.

Function in Food

Aspartame is a nutritive, non-cariogenic intense sweetener. The sweetness potency of aspartame is approximately 200 times that of sucrose. Additionally, it can act as a flavour enhancer, most noticeably with fruit flavours. Aspartame is normally digested in the body and has the same caloric value of 4 kcal/g as sugar, but, since considerably less is used, very few calories are added to the product. Aspartame is non-laxative and suitable for diabetics.

Benefits

Aspartame provides a clean sweetness, which can be slightly delayed and lasting. In terms of sweetness quality, it can be successfully combined with other intense sweeteners and with carbohydrate sweeteners. Aspartame shows synergistic behaviour with acesulfame K, which leads to a more sugar-like taste (masking the lingering sweetness of aspartame) and additional sweetness enhancement.

Limitations

Aspartame is permitted under the European Sweetener Directive 94/35/EC and its following amendments 96/83/EC and 2003/115/EC for use in more than 40 food applications.

Although aspartame is relatively stable in dry form, pH, temperature and time are very important factors affecting its stability in solution. The maximum stability of aspartame is obtained at pH 4.2–4.3. The hydrolysis between pH 3 and pH 5 can be limited under controlled temperatures. Below pH 3 and above pH 5, aspartame decreases rapidly even under ambient storage conditions. Therefore, aspartame is not very suitable for applications such as baked goods, since the manufacturing process involves exposure both to high pH levels and to high temperatures.

The Acceptable Daily Intake (ADI) value for aspartame was set at 40 mg/kg body weight according to JECFA. Owing to its phenylalanine content, persons suffering from the genetic disease phenylketonuria (PKU) must include the consumption of aspartame into their daily intake calculation. Therefore, a warning on aspartame-containing products is required in many countries (e.g. in the EU: "contains a source of phenylalanine").

Aspartame is approved for food use in most countries where intense sweeteners are used. Typical usage levels are up to 600 mg/litre in beverages and up to 5,500 mg/kg in chewing gums with no added sugar.

Typical Products

Aspartame is used in all fields of application of intense sweeteners. Key areas of application are: liquid and dry mix beverages, yoghurts (added after fermentation), ice cream and other dairy products, puddings, gelatins, confections, chewing gums, and table-top sweeteners. Owing to its synergistic characteristics, aspartame is often used in sweetener blends, e.g. beverages.

E952	Cyclamic acid and its salts

Sources

Cyclamic acid is manufactured by the sulphonation of cyclohexylamine and then converted to its sodium or calcium salt, which are the commercial forms of this sweetener.

Function in Food

Cyclamate is a potent, low-calorie sweetener. It is generally considered to be about 30 times as sweet as sucrose. Its primary commercial food uses are in soft drinks and table-top sweeteners.

Benefits

The sweet taste profile of cyclamate builds to a maximum more slowly than that of sucrose, but it also lingers for a longer time. It functions very effectively in combination with other potent sweeteners – in particular with saccharin, where that sweetener's bitter/metallic aftertaste is masked by cyclamate. When used in combination with saccharin, the normal ratio is 10:1

cyclamate:saccharin. This blend delivers a cost-effective, acceptable sweet taste profile. Cyclamate also synergises effectively with aspartame and acesulfame K, with commercial uses being in binary, tertiary and even quaternary blends with these sweeteners. It is particularly compatible with citrus flavours.

Sodium and calcium cyclamate are very soluble in water and solutions are stable at low pH to heat and to light.

Cyclamate is non-cariogenic and is able to mask bitter tastes effectively. Consequently, it is used in oral hygiene products and in liquid pharmaceutical preparations.

Cyclamate is used extensively in Asia. Europe consumes approximately 15% of the world supply.

Limitations

Cyclamate is included on the list of permitted sweeteners in Directive 94/35/EC as amended by Directive 2003/115/EC where it is permitted in a wide range of products with individual maxima.

An opinion of the EC Scientific Committee on Food in 2000 on cyclamic acid and its sodium and calcium salts established a new lower ADI of 0-7 mg/kg. In light of this, the maximum permitted level for cyclamate in soft drinks was reduced by Directive 2003/115/EC from 400 mg/l to 250 mg/l, and the use of the sweetener in sugar confectionery products is no longer permitted.

Typical Products

Soft drinks and table-top sweeteners

E953	Isomalt

Sources

Isomalt is a crystalline white substance with low hygroscopicity. It is a sugar replacer belonging to the group of polyols; more specifically, it is a disaccharide alcohol. Isomalt is produced by enzymic conversion of sucrose into isomaltulose, followed by hydrogenation into isomalt.

Isomalt is defined as a mixture of hydrogenated mono- and disaccharides whose principal components are the disaccharides 1,1-GPM dihydrate (1-0-α-D-Glucopyranosyl-D-mannitol dihydrate) and 1,6-GPS (6-0-α-D-Glucopyranosyl-D-sorbitol). Depending on the detailed composition of the saccharides and their intended application, different commercial variants in various particle sizes are available.

Function in Food

Isomalt is a sugar replacer. It has a sweetness of 0.5 to 0.6 times that of sugar.

The most important use of isomalt is in the production of confectionery and baked products with a reduced sugar or calorie claim. In these products it provides sweetness, bulk and texture while replacing sugar.

It is also used in products with a "tooth-friendly" claim since it is hardly fermented by plaque bacteria and reduces the formation of plaque by promoting remineralisation.

Benefits

The sweet taste of isomalt is similar to that of sugar but with a lower intensity. It is synergistic with most intense sweeteners.

Because only about half of the energy content of isomalt is actually utilised by the human body, the EU-approved caloric value is 2.4 kcal/g compared with 4 kcal/g for sugar (the figure for the USA and Canada is 2.0 kcal/g). It has a further benefit in that it has little effect on blood sugar levels.

Unlike some other polyols, it does not give a cooling effect in the mouth and it dissolves slowly so that sweets last longer.

Isomalt is not hygroscopic and does not participate in the Maillard (browning) reaction. These characteristics can be used to advantage in controlling moisture uptake during shelf-life and controlling colour development during cooking.

Limitations

In some countries isomalt is classed as a food, but in the EU it is regulated as an additive. Its use is covered by Directive 94/35/EC, as amended, when used as a sweetener, and by Directive 95/2/EC when used in other functions.

As with other polyols, if a foodstuff contains more than 10% isomalt, the product must be labelled "excessive consumption may produce laxative effects".

Typical Products

Hard and soft candies, chewing gum, chocolate products, baked goods, ice cream, jam, compressed tablets and coated products.

E954	Sodium saccharin

Sources

Sodium saccharin is a white crystalline powder synthesised from petroleum starting materials. It has been used as a sweetener for over 100 years.

Function in Food

Saccharin is an intense sweetener used to replace sugar in reduced-calorie products.

Benefits

Saccharin is approximately 450 times sweeter than sugar. It is stable to a range of processing conditions. It is not metabolised by the human body or by the bacteria that cause dental caries. It is produced in large quantities and is cost-effective in use. It is synergistic with other intense sweeteners.

Limitations

Saccharin is approved for use in over 100 countries. In the EU this approval is under the sweeteners Directive, 94/35/EC as amended, which permits its use in a range of products with individual limits in each case.

Saccharin has a bitter/metallic aftertaste.

Typical Products

Used in low- or reduced-calorie products, including soft drinks, jam, baked goods, canned fruit, dessert toppings, salad dressings and table-top sweeteners. It is also used in cosmetic products, pharmaceuticals and vitamin preparations.

E955	Sucralose

Sources

Sucralose is produced by the selective chlorination of three of the hydroxyl groups of sucrose to produce 1,6-dichloro-1,6-dideoxy-β-D-fructofuranosyl-4-chloro-4-deoxy-α-D-galactopyranoside.

It is freely soluble in water, methanol and ethanol and, is slightly soluble in ethyl acetate.

Function in Food

An intense sweetener.

Benefits

As sucralose is extremely soluble in water and at low temperatures, this makes it is easy to incorporate into foods and beverages.

One of the major technical advantages of sucralose is its ability to withstand high-temperature food processing and long-term storage even when used in low-pH products such as carbonated soft drinks.

Limitations

It is permitted under the Annex of Directive 2003/115/EC which amends Council Directive 94/35/EC on sweeteners for use in foodstuffs. It is permitted in a number of non-alcoholic drinks, desserts and similar products, confectionery and miscellaneous foodstuffs, all with specified limits.

Commission Directive 2006/128/EC which amends Directive 95/31/EC laying down specific criteria of purity concerning sweeteners for use in foodstuffs, specifies purity criteria for sucralose which must be met.

The acceptable daily intake (ADI) for sucralose is given by the Joint FAO/WHO Expert Committee on Food Additives as being between 0-15 mg/kg body weight.

Typical Products

Beverages and baked goods.

E957	Thaumatin

Sources

Thaumatin is a protein contained in the fruit of the plant *Thaumatococcus danielli*. Fruits are harvested from the green belts of Africa and part-processed to remove a section of the fruit known to contain thaumatin. Final processing in the UK is based on water extraction, ultrafiltration and freeze drying to produce thaumatin.

Function in Food

Thaumatin is a naturally sweet protein, approximately 2,500 times sweeter than sugar, and is used at very low levels, typically 0.5–3.0 ppm (below sweetness threshold), for its flavouring properties. It can mask unpleasant tastes, and synergise with other flavourings, sweeteners and flavour enhancers to improve both the taste and mouthfeel of a wide range of products.

Benefits

Thaumatin can mask bitterness and unpleasant aftertastes from soya, intense sweeteners, vitamins, minerals and herbs. It can reduce the off-notes arising during manufacture and storage of food and beverages, especially noted in citrus flavours. Thaumatin synergises with other ingredients to improve taste, whilst allowing a reduction in their levels. It also has the advantage of improving mouthfeel in low-fat or low-calorie products.

Limitations

Thaumatin complies with the requirements of article 1.2(c) of Directive 88/388/EEC on flavourings, as a flavouring preparation, and can be used for its flavouring properties without restriction in a wide variety of food products. It is permitted under both the sweeteners Directive, 94/35/EC and the miscellaneous additives Directive, 95/2/EC and its amendment 98/72/EC. In the former it is permitted as a sweetener in a limited range of products; in the latter it is permitted only in chewing gum, water-based flavoured non-alcoholic drinks and desserts.

Typical Products

Beverages, ice cream and desserts, chewing gum, low-sugar confectionery, savoury products, fortified foods, vitamin/mineral tablets and pharmaceuticals.

E959   Neohesperidine DC (NHDC)

Sources

Neohesperidine DC is prepared from the waste material from citrus processing.

Function in Food

NHDC is an intense sweetener, about 1,000 to 1,800 times more intense than sugar. It is used in very small amounts to enhance sweet taste and fruit flavours and to increase mouthfeel.

Benefits

NHDC is stable at high temperatures and has a long ambient shelf-life both as a powder and in aqueous solution.

Limitations

NHDC is permitted under both the sweeteners Directive 94/35/EC and the miscellaneous additives Directive 95/2/EC and its amendments in a number of products, with individual specified maxima. It has a distinctive taste and is unsuitable for use as the sole sweetener in products that do not require a liquorice taste.

Typical Products

Chewing gum, soft drinks, dairy products, ice cream and desserts.

E962	Salt of aspartame-acesulfame

Sources

The salt is prepared by heating an approximately 2:1 ratio (w:w) of aspartame and acesulfame K in solution at acidic pH and allowing crystallization to occur. The potassium and moisture are eliminated.

It is sparingly soluble in water and slightly soluble in ethanol.

Function in Food

An intense sweetener.

Benefits

The product is more stable than aspartame alone.

Limitations

It is permitted under the Annex of Directive 2003/115/EC which amends Council Directive 94/35/EC on sweeteners for use in foodstuffs. It is permitted in a number of foodstuffs such as non-alcoholic drinks, desserts and similar products, confectionery and miscellaneous foodstuffs, all up to specified limits.

The maximum usable doses for the salt of aspartame-acesulfame are derived from the maximum usable doses for its constituent parts, aspartame (E951) and acesulfame-K (E950). Hence, the maximum usable doses for both aspartame (E951) and acesulfame-K (E950) must not be exceeded by use of the salt aspartame-acesulfame, either alone or in combination with E950 or E951.

The acceptable daily intake (ADI) for the salt of aspartame-acesulfame is given by the Joint FAO/WHO Expert Committee on Food Additives as being between 0-15 mg/kg body weight.

A table top-sweetener containing the salt of aspartame and acesulfame must be labelled with statement to say 'contains a source of phenylalanine'.

Typical Products

Beverages, dairy products, table-top sweeteners and confectionery (e.g. chewing gum, hard candy).

E965	Maltitol

Sources

Maltitol and maltitol syrups are manufactured by hydrogenation of maltose and maltose/glucose syrup followed, for the crystalline maltitol, by a crystallisation and drying step. Maltitol syrups may be referred to as hydrogenated starch hydrolysate.

Function in Food

Maltitol exists as a pure crystalline material and as aqueous solutions having a dry matter content of 75%. It is a nutritive sweetener and replaces sucrose and glucose syrups, for bulk, texture and sweetness, in sugar-free confectionery products such as chocolate, chewing gum and hard-boiled, soft and chewy candies. Crystalline maltitol allows a very crunchy-chewing gum coating and helps control texture and flexibility, extending storage stability of chewing gum. Maltitol syrup acts as a plasticiser in chewing gum to give a more stable and softer texture. Pure crystalline maltitol is 90% as sweet as sucrose; the sweetening power of maltitol syrups ranges from 60 to 85%.

Benefits

Maltitol is not used as a food source by the bacteria that cause dental caries. It is not fully metabolised by the body and has been allocated a calorific value of 2.4 kcal/g in Europe and 2.1 kcal/g (3.0 kcal/g for maltitol syrups) in the USA. It does not participate in the Maillard reaction so it does not go brown on cooking.

Limitations

Maltitol is covered by both the sweeteners Directive, 94/35/EC as amended and the miscellaneous additives Directive 95/2/EC. In the latter, it is permitted in Annex IV for use in foods in general for purposes other than sweetening. As with all the polyols and some sources of dietary fibres, excessive consumption of maltitol can have a laxative effect, and products containing maltitol have to be labelled to this effect.

Typical Products

Sugar-free confectionery products such as chewing gum, chocolate, pastilles, gums, and hard-boiled, soft and chewy candies.

E966	Lactitol

Sources

Lactitol is a sugar alcohol produced by catalytic hydrogenation of lactose. Lactitol exists in three forms – dihydrate, monohydrate and anhydrous. The difference is in the amount of crystal bound water.

Function in Food

Lactitol is used as bulk sweetener in sugar-free, sugar-reduced and low-calorie foods, where it replaces sugar.

In processed meat, such as cooked ham, lactitol is used as a colour stabiliser. In surimi, lactitol acts as a cryoprotectant, and prevents denaturation of fish protein during freezing.

Lactitol can also be used as a prebiotic in all kinds of functional food, such as yoghurts and bakery products. In the colon, lactitol can be fermented by beneficial bacteria such as bifidobacteria and *Lactobacillus* spp.

Benefits

Lactitol is a disaccharide with similar physical properties to sucrose, but it has been determined to have a reduced calorific value (2.4 kcal/g in the EU, 2.0 kcal/g in the USA). However, it is only 30–40% as sweet as sugar. It does not participate in the Maillard reaction, so it does not contribute to browning on cooking, and it is not hygroscopic.

Like all polyols, lactitol is not fermented by mouth bacteria that cause dental caries. Lactitol is also metabolised independently of insulin, does not cause blood glucose levels to rise, and is suitable for diabetics.

Limitations

Lactitol is included in both the sweeteners Directive, 94/35/EC as amended and the miscellaneous additives Directive 95/2/EC. In the latter, it is permitted in Annex IV for use in foods in general for purposes other than sweetening.

As with all polyols, if a foodstuff contains more than 10% added lactitol, the product must be labelled "excessive consumption may produce laxative effects."

Typical Products

Products with energy-reduced, no-added-sugar, dietetic or tooth-friendly claims, including confectionery, baked goods, ice cream, chewing gum and jam. Lactitol is also used in ham.

E967	Xylitol

Sources

Xylitol is produced by the catalytic hydrogenation of xylose (wood sugar), which can be obtained from the xylan-rich hemicellulose portion of trees and plants. Xylitol is a natural constituent of many fruits and vegetables at levels of less than 1%, and the human body produces 5–15 g of xylitol per day during the metabolism of glucose.

Function in Food

Xylitol is principally used as a non-fermentable bulk sweetener in foods and oral hygiene products. In addition to its use as a sweetener, xylitol is also used as a humectant, as a masking agent for other ingredients, and as an energy source in intravenous products.

Benefits

Xylitol has a similar sweetness profile to that of sucrose, with no discernible aftertaste. In addition, xylitol has a distinct cooling effect in the mouth due to its negative heat of solution (the greatest of all the sweeteners).

Xylitol is suitable for use in diabetic foods, as it is metabolised independently of insulin and does not affect blood glucose levels following consumption.

Xylitol resists fermentation by oral bacteria and inhibits the growth of *Streptococcus mutans*, the organism most responsible for dental caries. The ability of xylitol to inhibit the development of new caries has been demonstrated in numerous clinical and field studies.

Limitations

Xylitol is approved within the EU under the Directive on sweeteners for use in foodstuffs (94/35/EC), and can be used to *quantum satis* in the applications specified. It is also permitted in foodstuffs, frozen fish and liqueurs for purposes other than sweetening, according to Directive 95/2/EC.

Xylitol is well tolerated but, as with other polyols, excessive consumption can cause laxative effects. The EC Scientific Committee on Food concluded that daily consumption of less than 20 g of polyols was unlikely to cause laxative effects except in sensitive individuals. However, as with all polyols, if a foodstuff contains more than 10% added xylitol, the product must be labelled "excessive consumption may produce laxative effects."

Typical Products

Xylitol is used in chewing gum, mints and gum-arabic pastilles, and other confectionery. It is also used in toothpaste, mouthwash and other dental speciality products, and as an excipient in pharmaceutical products.

E968	Erythritol

Sources

The starting material is a substrate obtained by the enzymatic hydrolysis of starch or sucrose. The glucose is then fermented by an osmophilic yeast *Moniliella pollinis* or *Trichosporonoides megachiliensis*. Erythritol is soluble in water and is slightly soluble in ethanol.

Function in Food

It can be used as a sweetener like the other currently permitted polyols. As well as use as a sweetener, it can act as a flavour enhancer, carrier, humectant, stabiliser, thickener, bulking agent, and sequestrant.

Benefits

Erythritol is permitted as a sweetener according to Directive 94/35/EC as amended by Directive 2006/52/EC. It may be used to *quantum satis* in a number of specified foods, including desserts and similar products, confectionery and food supplements as well as in table-top sweeteners.

It is listed in Annex IV to the miscellaneous additives Directive 95/2/EC, as amended by Directive 2006/52/EC, where it has been approved for use in foods for purposes other than sweetening to *quantum satis* in foodstuffs in general (except drinks and foodstuffs in which generally permitted additives are not permitted, e.g. unprocessed foodstuffs, honey, non-emulsified oils and fats of animal or vegetable origin and butter etc). It is also permitted in frozen and deep-frozen unprocessed fish, crustaceans, molluscs and cephalopods and in liqueurs.

It is listed as a permitted carrier and carrier solvent for food additives under Annex V to Directive 95/2/EC, as amended by Directive 2006/52/EC.

It is stable to heat and is non-hygroscopic.

Limitations
Erythritol has a laxative effect, but at a higher dose than other polyols.

Typical Products
Confectionery, desserts, food supplements, liqueurs.

E999   Quillaia extract

Sources
Quillaia extract is an aqueous extract of the bark of the tree Quillaia saponaria Molina. The solution is dried to provide a light brown powder, which is odourless but has an acrid, astringent taste. The active constituents are saponins, which are also present in sarsparilla, liquorice and yucca.

Function in Food
Quillaia extract is used to provide a stable foaming head on soft drinks such as ginger beer and cream soda.

Benefits
Quillaia extract provides a very stable foam and, at the levels used, is colourless and tasteless.

Limitations
According to Directive 95/2/EC as amended, quillaia is permitted only in non-alcoholic flavoured drinks and cider other than 'cidre bouché' to a maximum of 200 mg/litre measured as the solid extract.

Typical Products
Ginger beer and cream soda.

E1103   Invertase

Sources
Invertase is an enzyme that is normally present in human saliva. Industrially it is produced by submerged fermentation of yeast, from which it is separated and purified.

Function in Food
Invertase is used to produce invert sugar (a mixture of glucose and fructose) from sucrose.

Benefits
Invertase is used industrially to make invert syrup (golden syrup) from solutions of beet or cane sugar. It is also used in products with soft centres; the paste for the centre can be made with a firm texture but with the addition of invertase so that it softens after the assembly of the sweet but before consumption. The optimum pH is between 4.5 and 5.5 and the enzyme can work in liquid phases containing as much as 75% sucrose.

Limitations
Invertase is a generally permitted additive under Directive 98/72/EC.

Typical Products
Invertase is used in a range of confectionery products with soft or liquid centres.

E1105   Lysozyme

Sources
Lysozyme is an enzyme, extracted and purified from hen egg albumen.

Function in Food
Lysozyme is used to inhibit growth of the bacteria in hard cheese, which cause "late blowing".

Benefits
Lysozyme can be used in cheese in place of nitrate.

Limitations
Lysozyme is permitted only in ripened cheese (but the amount is limited only by Good Manufacturing Practice) and wine, according to Annex III part C of Directive 95/2/EC as amended by Directive 2003/114/EC.

E1200	Polydextrose

Sources

Polydextrose is prepared by a vacuum melt process involving polycondensation of glucose in the presence of small amounts of sorbitol and an acid. The final product of this reaction is a weakly acidic water-soluble polymer that contains minor amounts of bound sorbitol and acid. The polymer is then subjected to various clean-up procedures to produce several grades of polydextrose.

Function in Food

Polydextrose is a low-calorie bulking agent that is used as a partial replacement of sugars and/or fats whilst maintaining texture and mouthfeel. Polydextrose can be used to stabilise foods by preventing sugar and polyol crystallisation, e.g. in hard candies. Polydextrose functions as a humectant and retards the loss of moisture in baked goods, which helps protect against staling.

Benefits

Polydextrose is not sweet and can be used for both sweet and savoury products. It is only partially metabolised by bacteria in the large intestine and has been ascribed a caloric value of 1 kcal/g. Because it is digested in a similar way to dietary fibre and its metabolism does not involve insulin, polydextrose can be used in products designed for diabetic and low- glycaemic diets. Polydextrose is not fermented by mouth bacteria and will not promote dental caries.

Limitations

Polydextrose is included in Annex I of Directive 95/2/EC as a generally permitted additive.

Typical Products

Polydextrose is used in no-added-sugar, energy-reduced or dietetic products, including chocolate, hard candy, frozen dairy desserts, baked goods, fruit spreads and fillings, surimi, and beverages. In pharmaceutical preparations, solutions of polydextrose can be used as binders in wet granulation processes. Polydextrose may also be used in conjunction with other materials as a film and tablet-coating agent.

E1201   Polyvinylpyrrolidone

Sources

Polyvinylpyrollidone is made by a multistage synthesis starting from butan 1,3 diol, followed by purification.

Function in Food

Polyvinylpyrollidone is used to help tablets break up in water.

Benefits

Polyvinylpyrollidone is water-soluble and is used in tablet coatings to increase the penetration of water into the tablet.

Limitations

Polyvinylpyrrolidone is permitted only in tableted food supplements and in sweetener preparations according to Directive 95/2/EC, as amended by Directive 2006/52/EC.

E1202   Polyvinylpolypyrrolidone

Sources

Polyvinylpolypyrollidone is a white powder made by crosslinking polyvinylpyrollidone followed by purification.

Function in Food

Polyvinylpolypyrrolidone is used to help tablets break up in water. It is also used as a processing aid in the treatment of wine because it strongly binds tannins.

Benefits

Polyvinylpolypyrollidone is water-absorbent and swells in water, but is not water-soluble. When included in a tablet, this swelling in water is used as a means of making the tablet break up.

Limitations

Polyvinylpolypyrrolidone is permitted only in tableted food supplements and in sweetener preparations according to Directive 95/2/EC, as amended by Directive 2006/52/EC.

E1204   Pullulan

Sources
Pullulan is a linear, neutral glucan which consists primarily of maltotriose units connected by α-1-6 glycosidic bonds. It is produced by fermentation from a food grade hydrolysed starch using a non-toxin producing strain of the fungus *Aureobasidium pullulans*. Once the fermentation is complete, the fungal cells are removed by microfilitration, the filtrate is heat-sterilised and pigments and other impurities are removed by adsorption and ion exchange chromatography.

Function in Food
Glazing agent, film-forming agent, thickener.

Benefits
The α-1-6 linkages are thought to be responsible for the structural flexibility and solubility of pullulan which give it distinct film and fibre-forming characteristics that are not shown by other polysaccharides.

Limitations
It is permitted under Annex IV to Directive 95/2/EC, as amended by Directive 2006/52/EC but is acceptable for use only in the coating of food supplements that are in capsule/tablet form and breath freshening microsweets in film form to *quantum satis*.

The acceptable daily intake (ADI) for Pullulan is given by the Joint FAO/WHO Expert Committee on Food Additives as being 'not specified'.

Typical Products
Food supplements (capsule/tablet form), breath freshening microsweets (in the form of films).

E1404   Oxidised starch

Sources
Native starches are oxidised by treating an aqueous starch suspension with sodium hypochlorite.

Function in Food

Oxidised starches provide soft gels, which exhibit greater stability in high-sugar systems and greater resistance to shrinkage. They are also used to improve the adhesion properties of batters.

Benefits

Oxidised starches can be used at higher dosage rates than their parent native starches, thus increasing the range of textures achievable in gum confectionery. A further benefit is increased and improved shelf-life. Oxidised starches, used in adhesion batters, can improve the visual and eating quality of battered foods, owing to a more consistent coverage of the substrate.

Limitations

Oxidised starches are generally permitted additives under Annex I of Directive 95/2/EC.

Typical Products

Oxidised starches are used in gum confections and lozenges, dairy products, and batters and breadings as coatings for poultry, meat, fish and vegetables.

E1410   Monostarch phosphate

Sources

Native starches are phosphorylated to produce these modified starches (also referred to as stabilised starches), where only one starch hydroxyl group is involved in the starch-phosphate linkage. Typical reagents include ortho-phosphoric acid, sodium or potassium ortho-phosphate, or sodium tripolyphosphate.

Function in Food

These modified starches are used as freeze-thaw-stable thickeners for simple processes. For greater process tolerance, cross-linked starches are required. Starch phosphates also exhibit good emulsifying properties. Pregelatinised starch phosphates are used as thickeners in dry mix puddings and as binders in bakery products.

Benefits

Monostarch phosphates improve the product quality and shelf-life stability of frozen foods. Product quality is also improved in salad dressings when they are used as emulsion stabilisers. Their incorporation into baked goods can improve moisture retention, which enhances eating quality and extends shelf-life.

Limitations
Monostarch phosphates are generally permitted additives under Annex I of Directive 95/2/EC.

Distarch phosphates (cross-linked starches) are used more extensively than monostarch phosphates in the food industry as thickening agents.

Typical Products
Frozen gravies, pie fillings, and salad dressings.

E1412   Distarch phosphate

Sources
Native starches are cross-linked by reacting an aqueous starch slurry with reagents such as phosphorous oxychloride or sodium trimetaphosphate. A range of modified starches is available with different levels of cross-linking.

Function in Food
These modified starches are thickeners, which provide short, salve-like textures in processed foods. The choice of starch will depend on processing conditions – heat, acid and shear – in order to achieve adequate granule swelling for optimal viscosity development. These starches produce pastes with fast meltaway.

Benefits
Cross-linked starches offer heat-, acid- and shear-stability. They provide stable viscosity in a wide range of heat processes where native starches would break down with a significant loss of viscosity. Consequently, they can be used at lower dosage rates than their parent native starches.

Limitations
Distarch phosphates are generally permitted additives under Annex I of Directive 95/2/EC.

Distarch phosphates are not recommended in chilled or frozen applications. Such products would require both cross-linked and stabilised starches, such as acetylated distarch adipates or hydroxypropylated distarch phosphates.

Typical Products
Bottled sauces, salad dressings, dry-mix puddings and baked goods.

E1413   Phosphated distarch phosphate

Sources
Native starches are modified by a combination of treatments, as for monostarch phosphates and distarch phosphates, on an aqueous slurry. Typical reagents include ortho-phosphoric acid, sodium or potassium ortho-phosphate, sodium tripolyphosphate with phosphorous oxychloride or sodium trimetaphosphate.

Function in Food
These modified starches are used as freeze-thaw-stable thickeners. The choice of starch will depend on processing conditions – heat, acid and shear – in order to achieve adequate granule swelling for optimal viscosity and development. These starches produce pastes with fast meltaway.

Benefits
Cross-linked and stabilised starches improve the product quality and shelf-life stability of foods. These modified starches offer greater process stability and low-temperature storage stability than their parent native starches and consequently can be used at a lower dosage rate.

Limitations
Phosphated distarch phosphates are generally permitted additives under Annex I of Directive 95/2/EC.

Typical Products
Bottled sauces, frozen gravies and pie fillings.

E1414   Acetylated distarch phosphate

Sources
Native starches are cross-linked and stabilised by reacting an aqueous slurry with reagents such as phosphorous oxychloride or sodium trimetaphosphate combined with acetic anhydride or vinyl acetate. A range of

modified starches is available with different levels of cross-linking and stabilisation.

Function in Food

These modified starches are thickeners and stabilisers, which are designed to maintain granular integrity throughout processing to provide short, salve-like textures in processed foods. The textural qualities are retained even after the processed foods have been chilled or frozen. The choice of starch will depend on processing conditions – heat, acid and shear – in order to achieve adequate granule swelling for optimal viscosity development. These starches produce pastes with fast meltaway.

Benefits

Acetylated distarch phosphates provide stable viscosity in a wide range of heat processes where native starches would break down with a significant loss in viscosity. They also increase shelf-life by providing low-temperature stability for chilled and frozen products. In such applications, native starches, particularly amylose-containing starches, would limit shelf-life, owing to retrogradation or syneresis.

Limitations

Acetylated distarch phosphates are generally permitted additives under Annex I of Directive 95/2/EC.

Typical Products

Soups, sauces, dairy products, fruit fillings, pet foods, and chilled and frozen recipe dishes.

E1420   Acetylated starch

Sources

Acetylated starches are produced from native starches by reacting an aqueous slurry with acetic anhydride or vinyl acetate.

Function in Food

Starch acetates, when cooked in water, rapidly develop a stable viscosity with a reduced tendency to set back or retrograde on cooling. Paste clarity is also improved.

Benefits

Acetylated starches are easier to cook, owing to lowering of the gelatinisation temperature. This is a particular benefit for high-amylose starches, which are difficult to cook at ambient pressure, whereas acetylation renders them dispersable under such conditions. The high-solids environment of confectionery products limits the scope for viscosity control unless an easy-to-cook thickener, such as acetylated starch, is used. Shelf-life stability, particularly in chilled and frozen products, is extended with starch acetates.

Limitations

Acetylated starches are generally permitted additives under Annex I of Directive 95/2/EC.

Acetylated starches often cause curdling in dairy products. This has been attributed to the instability of the acetate linkage in high protein concentrations. In applications such as these, hydroxypropylated starches are more compatible with milk proteins and are therefore recommended for greater stability. Stabilised and cross-linked starches, such as acetylated distarch phosphates or adipates, are more suitable for a wider range of heat-processed foods than starch acetates. Acetylated starches are not as freeze-thaw-stable as hydroxypropylated starches.

Typical Products

Batters and breadings, snacks, cereals and confectionery products.

E1422   Acetylated distarch adipate

Sources

Native starches are esterified with acetic anhydride and adipic anhydride to produce acetylated distarch adipates. A range of modified starches is available with different levels of cross-linking and stabilisation.

Function in Food

These cross-linked and stabilised starches are the most commonly used starch-based thickeners and stabilisers in processed foods. They are designed to provide viscosity stability and tolerance to heat, acid, shear and low-temperature storage. These starches produce pastes with fast meltaway.

Benefits

These modified starches allow the manufacture and distribution of high-quality processed foods. Cooked starch pastes have a higher cold viscosity, owing to their ability to maintain starch structure after processing. Heat penetration is enhanced with more highly cross-linked starches. Shelf-life is extended, particularly in chilled and frozen products.

Limitations

Acetylated distarch adipates are generally permitted additives under Annex I of Directive 95/2/EC.

Acetylated starches often cause curdling in dairy products. This has been attributed to the instability of the acetate linkage in high protein concentrations. In applications such as these, hydroxypropylated starches are more compatible with milk proteins and are therefore recommended for greater stability. Acetylated starches are not as freeze-thaw-stable as hydroxypropylated starches.

Typical Products

Gravies, soups, sauces, mayonnaise and dressings, sweet and savoury fillings, fruit preparations, dairy products, and chilled and frozen recipe dishes.

E1440   Hydroxypropyl starch

Sources

Native starches are reacted with propylene oxide to produce this range of ether-derivatised starches. They represent an alternative range of stabilised starches.

Function in Food

These modified starches bind water at lower temperatures than their parent native starches to texturise certain foods and introduce low-temperature stability. Hydroxypropyl starches are frequently further modified to increase their range of applications. The most common combination treatment is cross-linking, where cross-linked and stabilised starches are ideal in chilled and frozen foods.

Benefits

Hydroxypropylated starches are easier to cook than their parent native starches, owing to the reduction in gelatinisation temperature. This is ideal in low-moisture products or high-solids cooking, where competition for water makes it difficult to cook the starch fully and therefore develop maximum stable viscosity.

Limitations

Hydroxypropylated starches are generally permitted additives under Annex I of Directive 95/2/EC. These modified starches are not permitted in baby foods.

Typical Products

Meats, beverages and low-fat and low-calorie products.

E1442   Hydroxypropyl distarch phosphate

Sources

These modified starches are produced from the reaction of an aqueous starch suspension with a combination of propylene oxide and either sodium trimetaphosphate or phosphorous oxychloride. The latter two reagents are cross-linking agents, whilst the former stabilises the starch by etherification. A range of modified starches is available with different levels of cross-linking and stabilisation.

Function in Food

These cross-linked and stabilised starches are widely used as thickeners, stabilisers and mouthfeel enhancers. The choice of starch will depend on processing conditions – heat, acid and shear – in order to achieve adequate granule swelling for optimal viscosity development and stability. These modified starches also confer excellent low-temperature and freeze-thaw stability. They produce thick, rich, creamy pastes with excellent mouthfeel and cling.

Benefits

The mouthfeel of these starches improves the product aesthetics of low-fat/low-calorie or fat-free products. Their high viscosity allows lower usage rates than is the case with native starches, and they can give a number of processing benefits results from their rapid cooking and low fouling of process plant.

Limitations

Hydroxypropylated distarch phosphates are generally permitted additives under Annex I of Directive 95/2/EC. These modified starches are not permitted in baby foods.

Typical Products

Gravies, soups, sauces, mayonnaises and dressings, sweet and savoury fillings, fruit preparations, dairy products, chilled and frozen recipe dishes, meat and meat analogues, and pet foods.

E1450   Starch sodium octenylsuccinate

Sources

Native starches are modified by substitution using 1-octenylsuccinic anhydride (OSA) or succinic acid.

Function in Food

These modified starches are effective emulsion stabilisers, owing to the introduction of a hydrophobic moiety onto the starch polymer.

Benefits

Low-viscosity OSA-treated starches can be used at higher solids levels in spray-dried applications, which improves plant efficiency by reducing drying times. These modified starches have better film-forming properties, which results in better oxidation stability and therefore shelf-life stability. Emulsified products benefit from these properties, which improve product aesthetics and extend shelf-life. OSA-treated starches are also designed for low-temperature storage, giving temperature tolerance and flexibility in handling of flavour emulsions.

Limitations

Starch sodium octenylsuccinates are generally permitted additives under Annex I of Directive 95/2/EC.

Typical Products

Spray-dried flavours, beverage emulsions, emulsified sauces, mayonnaises and salad dressings.

E1451   Acetylated oxidised starch

Sources

Native starches are modified by oxidising agents, such as hypochlorite, followed by acetylating agents, such as acetic anhydride.

Function in Food
These modified starches are used as binding and gelling agents in confectionery. Oxidised starches have lower gelatinisation temperatures and hot viscosity, and improved paste clarity and low-temperature-storage stability. The combined acetylation treatment enhances these properties.

Benefits
These starches can be used as alternatives to gelatin and gum arabic in confectionery products.

Limitations
Acetylated oxidised starches are generally permitted additives under Directive 95/2/EC.

Typical Products
Soft sugar confectionery products.

E1452   Starch aluminium octenyl succinate

Sources
Starch aluminium octenyl succinate is the aluminium salt of the production of octenyl/succinic anhydride with starch.

Function in Food
Powdering agent.

Benefits
Starch aluminium octenyl succinate is an effective anticaking agent for use in vitamin and nutrient preparations.

Limitations
It is permitted under Annex IV to Directive 95/2/EC, as amended by Directive 2006/52/EC, for the use only in encapsulated vitamin preparations in food supplements up to 35 g/kg in food supplements.

Typical Products
It is used as a component of micro encapsulated vitamins and carotenoids.

E1505   Triethyl citrate

Sources

Triethyl citrate is made by reacting citric acid with ethanol. It is an odourless and colourless oily liquid.

Function in Food

Triethyl citrate is used to increase the rate at which rehydrated egg white powder forms a stable foam. It can also be used as an antifoaming agent, sequestrant, stabiliser or as a carrier solvent.

Limitations

Triethyl citrate is permitted according to Directive 95/2/EC, as amended by Directives 2003/114/EC and 2006/52/EC, in a limited number of foodstuffs, namely dried egg white to *quantum satis* and in flavourings up to specified limits.

E1517   Glyceryl diacetate (diacetin)

Sources

Glyceryl diacetate consists predominantly of a mixture of the 1,2- and 1,3-diacetates of glycerol, with minor amounts of the mono- and tri-ester. It is soluble in water and is miscible with ethanol.

Function in Food

Carrier solvent.

Limitations

It is permitted only for use in flavourings up to a specified level when in foodstuffs according to Directive 95/2/EC as amended by Directives 2003/114/EC and 2006/52/EC.

E1518   Glyceryl triacetate (triacetin)

Sources

Glyceryl triactetate is also known as triacetin. It is a colourless oily liquid prepared by the reaction of glycerol with acetic anhydride. It is sparingly soluble in water and soluble in ethanol.

Function in Food

Glyceryl triacetate is a hydrophobic liquid used as a lubricant in chewing gum.

Limitations

Glyceryl triacetate is only slightly soluble in water.

According to Directive 95/2/EC as last amended by Directive 2006/52/EC, it is permitted as a carrier and for use in chewing gum according to Good Manufacturing Practice. It is also permitted for use in flavourings up to specified limits in the final foodstuffs.

The acceptable daily intake (ADI) for Glyceryl triacetate is given by the Joint FAO/WHO Expert Committee on Food Additives as being 'not specified'.

E1519   Benzyl alcohol

Sources

Benzyl alcohol is a natural constituent of a number of plants. It occurs, for example, in some edible fruits (up to 5 mg/kg) and in green and black tea (1-30 and 1-15 mg/kg, respectively).

Function in Food

It is used as a carrier solvent for flavourings.

Limitations

It is permitted in flavourings for a limited number of foodstuffs: liqueurs, aromatised wines, aromatised wine-based drinks and aromatised wine-products cocktails as well as confectionery including chocolate and fine bakery wares up to specified levels under Annex IV of Directive 95/2/EC, as amended by 2003/114/EC.

Typical Products

Benzyl alcohol is added as a flavouring substance to some foods including chewing gum and beverages.

E1520   Propan-1,2-diol (propylene glycol)

Sources

Propane-1,2-diol, also known as propylene glycol, is produced by heating glycerol with sodium hydroxide, or by treating propylene with chlorinated water to form the chlorohydrin, followed by a further treatment with sodium carbonate solution to form the glycol.

Function in Food

It can be used as an anticaking agent, antifoaming agent, emulsifier, flour treatment agent, humectant, stabiliser, thickener, adjuvant and carrier solvent.

Limitations

Propane-1,2-diol is permitted under Annex V of Directive 95/2/EC as last amended by Directive 2006/52/EC as a carrier solvent for colours, emulsifiers, antioxidants and enzymes such that the maximum level occurring in the final foodstuff shall not exceed 1 g/kg and for use in flavourings up to specified limits in the final foodstuffs. It is permitted in flavourings up to 1g/l for use in beverages (excluding cream liqueurs).

The acceptable daily intake (ADI) for propylene glycol is given by the Joint FAO/WHO Expert Committee on Food Additives as being between 0-25 mg/kg body weight.

ABBREVIATIONS USED IN TEXT

ADI	Acceptable daily intake
ANZFA	Australia New Zealand Food Authority
ANZFSC	Australia New Zealand Food Standards Council
ATF	Bureau of Alcohol, Tobacco and Firearms
CCFA	Codex Committee on Food Additives
CCFAC	Codex Committee on Food Additives and Contaminants
CCCF	Codex Committee on Contaminants in Foods
CFSAN	Center for Food Safety and Applied Nutrition
COT	Committee on Toxicity of Chemicals in Food, Consumer Products and the Environment
EFSA	European Food Safety Authority
EU	European Union
FAC	Food Advisory Committee
FAO	Food and Agriculture Organization
FDA	Food and Drug Administration
FEMA	Flavor and Extract Manufacturers' Association
FFDCA	Federal Food, Drug and Cosmetic Act
FSA	Food Standards Agency
GMP	Good Manufacturing Practice
GRAS	Generally recognised as safe
GSFA	General Standard on Food Additives
INS	International Numbering System
JECFA	Joint FAO/WHO Expert Committee on Food Additives
MAFF	Ministry of Agriculture, Fisheries and Food
MERCOSUR	Treaty for the Organisation of a Southern Common Market
NAFTA	North American Free Trade Agreement
NOAEL	No observed adverse effect level
OECD	Organisation for Economic Co-operation and Development
SURE	Safety Universally Recognised list
SCF	Scientific Committee on Food
TVC	Total viable count
USDA	United States Department of Agriculture
WHO	World Health Organization
WTO	World Trade Organization

GLOSSARY

Acute toxicity: Adverse health effects induced by a single dose of a chemical

Adrenal phaeochromocytomas: Brown-staining tumour of the adrenal gland that produces catecholamines (e.g. adrenaline)

Atopic: A constitutional tendency to develop allergic reactions

Azo dye: A coloured organic chemical containing the azo grouping –N=N–

Bioassay: Toxicity test in animals, usually of 2-years' duration, in which the chronic toxicity and carcinogenicity of a test chemical are investigated

Caecal enlargement: Increase in weight of the caecum (beginning of the large intestine), often seen in rodent toxicity tests when large amounts of poorly digestible substances are administered orally

Calcium homeostasis: Maintenance of the balance between the loss of calcium from the bone via urine and faeces and resorption of calcium in the bone

Chelate: A complex in which metals are tightly bound to organic chemicals

Chronic toxicity: Adverse health effects induced by repeated exposure to a chemical over a period of several months or more

Dose–response relationship: Relationship between the amount of a chemical administered and the nature of the toxic effect induced

Embryogenesis:

Period of major organ development in the embryo

Endocrine disrupter:

Substance causing adverse health effects in an animal or its offspring secondary to an effect on the endocrine (hormonal) system

Epithelium:

Layers of cells covering the external surface of the body or lining hollow structures (e.g. gut) within the body

Gavage:

Administration of chemicals to animals by stomach tube

Haematological:

Pertaining to blood

Hyperplasia:

An increase in the number of cells, sometimes a precursor to tumours

Lymphocytes:

Type of white blood cell involved in immune reactions (e.g. fighting infections and allergic reactions)

Organogenesis:

Time period when the major organs are developing in the embryo and foetus

Osmotic diarrhoea:

Watery diarrhoea caused by the presence of large amounts of poorly digestible substances in the large intestine that draw water from inside the body into the intestinal lumen by osmosis

Pathological changes:

Undesirable changes to tissues and fluids in the body caused by disease or toxic chemicals

Pelvic nephrocalcinosis:

Deposition of calcium in the kidneys

Sensitisation:

Part of the allergic process in which an individual becomes hypersensitive to a substance and suffers an allergic reaction when next exposed to it

Stereoisomer

One of two or more compounds that differ only in their spatial arrangement of atoms

Urticaria:

A condition in which weals appear on the skin and itch intensely as a result of taking drugs or certain foods or as a reaction to the injection of serum, insect bites or the stings of plants (nettle rash)

INDEX